Lecture Notes in Engineering

Edited by C. A. Brebbia and S. A. Orszag

4

W. S. Venturini

Boundary Element Method in Geomechanics

Springer-Verlag
Berlin Heidelberg New York Tokyo 1983

Series Editors
C. A. Brebbia · S. A. Orszag

Consulting Editors
J. Argyris · K.-J. Bathe · A. S. Cakmak · J. Connor · R. McCrory
C. S. Desai · K.-P. Holz · F. A. Lecki · F. Pinder · A. R. S. Pont
J. H. Seinfeld · P. Silvester · W. Wunderlich · S. Yip

Author
W. S. Venturini
Departamento de Estruturas
Escola de Engenharia de São Carlos
Universidade de São Paulo
Av. Dr. Carlos Botelho, 1465
13560 – São Carlos – SP
Brazil

Formerly Southampton University, England

With 114 Figures

ISBN 3-540-12653-8 Springer-Verlag Berlin Heidelberg New York Tokyo
ISBN 0-387-12653-8 Springer-Verlag New York Heidelberg Berlin Tokyo

Library of Congress Cataloging in Publication Data
Venturini, W.S., 1948–
Boundary element method in geomechanics.
(Lecture notes in engineering ; 4)
Bibliography: p.
1. Soil mechanics. 2. Boundary value problems.
I. Title. II. Series
TA710.V427 1983   624.1'51'0151535   83-12388

This work is subject to copyright. All rights are reserved, whether the whole or part of the material is concerned, specifically those of translation, reprinting, re-use of illustrations, broadcasting, reproduction by photocopying machine or similar means, and storage in data banks.
Under § 54 of the German Copyright Law where copies are made for other than private use, a fee is payable to 'Verwertungsgesellschaft Wort', Munich.

© Springer-Verlag Berlin Heidelberg 1983
Printed in Germany

Printing and binding: Beltz Offsetdruck, Hemsbach/Bergstr.
2061/3020-543210

CONTENTS

NOTATION		VI
CHAPTER 1	INTRODUCTION	1
CHAPTER 2	MATERIAL BEHAVIOUR AND NUMERICAL TECHNIQUES	11
2.1	Introduction	11
2.2	Linear Elastic Material Problems	12
2.3	Nonlinear Elastic Material Problems	15
2.4	Inelastic Material Problems	19
2.5	Time-Dependent Problems	23
CHAPTER 3	BOUNDARY INTEGRAL EQUATIONS	27
3.1	Introduction	27
3.2	Governing Equations and Fundamental Solutions	27
3.3	Integral Equations	36
3.4	Body Force Problem	41
3.5	Prestress Force Problem	43
3.6	Temperature Shrinkage and Swelling	46
CHAPTER 4	BOUNDARY INTEGRAL EQUATIONS FOR COMPLETE PLANE STRAIN PROBLEMS	49
4.1	Introduction	49
4.2	Governing Equations and Fundamental Solutions	49
4.3	Integral Equations for Interior Points	55
4.4	Boundary Integral Equation	58
CHAPTER 5	BOUNDARY ELEMENT METHOD	60
5.1	Introduction	60
5.2	Discretization of the Integral Equations	61

	5.3	Subregions	73
	5.4	Traction Discontinuities	76
	5.5	Thin Subregions	79
	5.6	Solution Technique	82
	5.7	Practical Application of Boundary Element on Linear Problems	85
CHAPTER 6		NO-TENSION BOUNDARY ELEMENTS	95
	6.1	Introduction	95
	6.2	Rock Material Behaviour	95
	6.3	Method of Solution	97
	6.4	Application of No-Tension in Rock Mechanics	102
CHAPTER 7		DISCONTINUITY PROBLEMS	118
	7.1	Introduction	118
	7.2	Plane of Weakness	119
	7.3	Analysis of Discontinuity Problems	122
	7.4	Numerical Applications	127
CHAPTER 8		BOUNDARY ELEMENT TECHNIQUE FOR PLASTICITY PROBLEMS	135
	8.1	Introduction	135
	8.2	Elastoplastic Problems in One Dimension	135
	8.3	Theory of Plasticity for Continuum Problems	141
	8.4	Numerical Approach for the Plastic Solution	151
	8.5	Practical Applications in Geomechanics	152
CHAPTER 9		ELASTO/VISCOPLASTIC BOUNDARY ELEMENT APPROACH	165
	9.1	Introduction	165
	9.2	Time-Dependent Behaviour in One Dimension	165
	9.3	Elasto/Viscoplastic Constitutive Relations for Continuum Problems	172

9.4	Outline of the Solution Technique	175
9.5	Time Interval Selection and Convergence	176
9.6	Elasto/Viscoplastic Applications	179

CHAPTER 10	APPLICATIONS OF THE NONLINEAR BOUNDARY ELEMENT FORMULATION	189
10.1	Introduction	189
10.2	Strip Footing Problem	190
10.3	Slope Stability Analysis	199
10.4	Tunnelling Stress Analysis	206

CHAPTER 11	CONCLUSIONS	218
REFERENCES		227
APPENDICES		241

NOTATION

$\underset{\sim}{A}$	matrix of the final system of equations
b_i	body forces
$\underset{\sim}{B}$	body force vector
c	cohesion of the material
c_{ij}	coefficients of the free term
$\left.\begin{array}{l} D_{ijk} \\ S_{ijk} \\ \varepsilon^*_{\ell mk} \\ \sigma^*_{\ell mk} \\ E_{ijmk} \\ F_{ijmk} \end{array}\right\}$	components of tensors corresponding to derivatives of the fundamental solution.
E	modulus of elasticity
$\underset{\sim}{F}$	independent vector of the final system of equations
G	shear modulus
$\lvert G \rvert$	Jacobian (for boundary elements)
H'	slope of the uniaxial stress strain curve
$\left.\begin{array}{l} \underset{\sim}{H} \\ \underset{\sim}{G} \\ \underset{\sim}{D} \\ \underset{\sim}{E} \end{array}\right\}$	matrices for the boundary equation
$\left.\begin{array}{l} \underset{\sim}{H'} \\ \underset{\sim}{B'} \\ \underset{\sim}{D'} \\ \underset{\sim}{E'} \end{array}\right\}$	matrices for stress determination
$\underset{\sim}{I}$	unit matrix
$\lvert J \rvert$	Jacobian (for internal cells)
k	hardening parameter
$\underset{\sim}{M}$	elastic boundary solution; vectorial form
$\underset{\sim}{N}$	elastic stress solution; vectorial form

p_i	traction components
$\underset{\sim}{p}$	traction vector
$\underset{\sim}{P}$	nodal traction vector
p^*_{ij}	traction fundamental solution
$\underset{\sim}{p}^*$	matrix of fundamental tractions
q	field point inside the domain
Q	field point on the boundary
r	distance between the field and load points
s	load point inside the domain
S	load point on the boundary
t	time
t_m	overlay thickness
u_i	displacement components
$\underset{\sim}{u}$	displacement vector
$\underset{\sim}{U}$	nodal displacement vector
u^*_{ij}	displacement fundamental solution
$\underset{\sim}{u}^*$	matrix of fundamental displacements
x_i	coordinates of the global system
$\bar{x}_i$	coordinates of the local system
$\bar{\gamma}$	unit weight
γ	viscosity parameter
Γ	boundary of the body
δ_{ij}	Kronecker delta
$\delta(s,q)$	Dirac delta function
ε_{ij}	strain components
ε^e_{ij}	elastic strain components
ε^o_{ij}	initial strain components
$\dot{\varepsilon}_{ij}$	strain rate components
$\dot{\varepsilon}^{vp}_{ij}$	viscoplastic strain rate components
η_i	direction cosines

ν	Poisson's ratio
ξ	homogeneous coordinate over a boundary element
ξ_i	homogeneous triangular coordinates
σ_{ij}	stress components
σ_{ij}^o	initial stress components
σ_{ij}^e	elastic stress components
ϕ	angle of internal friction
$\underline{\psi}, \underline{\Phi}$	interpolation functions

CHAPTER 1

INTRODUCTION

Numerical techniques for solving many problems in continuum mechanics have experienced a tremendous growth in the last twenty years due to the development of large high speed computers. In particular, geomechanical stress analysis can now be modelled within a more realistic context.

In spite of the fact that many applications in geomechanics are still being carried out applying linear theories, soil and rock materials have been demonstrated experimentally to be physically nonlinear. Soils do not recover their initial state after removal of temporary loads and rock does not deform in proportion to the loads applied. The search for a unified theory to model the real response of these materials is impossible due to the complexities involved in each case. Realistic solutions in geomechanical analysis must be provided by considering that material properties vary from point to point, in addition to other significant features such as non-homogeneous media, in situ stress condition, type of loading, time effects and discontinuities.

A possible alternative to tackle such a problem is to introduce some simplified assumptions which at least can provide an approximate solution in each case. The validity or accuracy of the final solution obtained is always dependent upon the approach adopted. As a consequence, the choice of a reliable theory for each particular problem is another difficult decision which should be taken by the analyst in geomechanical stress analysis.

In rock mechanics stress analyses, mainly those related to excavations, material responses under tensile and compressive stresses have to be distinguished either as a consequence of the original fissured state or due to blasting during the excavation process. Then, the presence of this fissured state is an imperative that the rock cannot withstand tension, though the individual blocks may carry some tensile stresses. Such a behaviour should be taken into account to provide realistic stress determination in the rock medium as well as the final loading in the additional structures, which may be necessary at the construction time or at the utilization stage.

The anisotropic effects are another important aspect to be taken into consideration in rock mechanics. This is not only due to usual material properties being different in all directions, but also due to fissured state of the rock which may contain cracks filled or not with other softer materials. Also, slidings and relative displacements along preexisting planes of weakness or discontinuities already present in the rock media can modify significantly the final solutions.

An important aspect to be considered in soil and rock mechanics is the highly nonlinear stress-strain relationship which gives permanent deformation on unloading. For different reasons soil and rock have been considered to behave as elastoplastic materials with or without hardening or softening effects. Based upon the work of Drucker et al (10,11,12), plastic concepts were extended to soil material, and limit theorems for the prediction of limit loads were also established. Other models have been developed in the plasticity theory context trying to produce better correlations between theoretical and practical results. In particular, work carried out at Cambridge University (13) has been recognised throughout the engineering research world as an improvement on the establishment of realistic material responses.

All plastic theories for any material which are or are not in this context are based upon assumptions that are totally independent of time. They consider that permanent and irreversible deformations are set up instantaneously. However, as is observed in real soil and rock materials, the influence of time is an important aspect to be taken into account in any geomechanical constitutive law.

The inclusion of time-dependent effects, which are usually spoken of under the general title of creep, is of great importance in rock mechanics. Many criteria (14) to deal with the viscous deformation have been developed since the first studies in this area, carried out by Gribbs (15). Depending on the stress level, viscoelasticity or primary creep can be assumed. In this case, a final deformation state is reached and the recovery of the initial state is possible when the load is removed. However, in some rocks, specially the salt type, the stress level usually reaches a certain creep yield limit which leads to the secondary creep, where deformations are no longer recoverable and their rate can be taken as constant.

Time-dependent response for soil and rock materials can also be obtained by adopting a viscoplastic model due to Perzyna (16), in which the yield stress level is monitored by a well established plastic criterion, and the viscoplastic effects computed according to the same yield function since associated plasticity is assumed.

This model is extremely versatile for geomechanical applications; it not only models simple viscoplastic responses but can also model other more complex viscous laws by adopting a combination of two or more units in parallel or in series. In particular, the Kelvin and Maxwell units, which are useful in many rock and soil problems for the viscoelastic and creep representations, can be easily obtained by either appropriate combinations or particularizations of the original model.

Another characteristic of the Perzyna's formulation is that it can be applied to model pure plastic behaviour. As the theory takes into account a well established yield surface, the plastic solution is obtained when the loads are applied in small increments followed by stationary conditions.

The usefulness of numerical techniques in solving the above mentioned geomechanical problems has been fully recognized by geotechnical engineers. Quite generally these techniques can be divided into three main categories: finite difference, finite element and boundary element methods.

The finite difference technique was the first numerical approach formulated on mathematical bases which has been applied in continuum mechanics stress analysis. The method started as a numerical technique after Southwell (19) had presented his relaxation method, although no computer automatization was possible at that time. The finite difference remained until the computer "era", when the technique spread widely to solve problems in the context of continuum mechanics; it has also been particularly suitable for certain classes of applications such as shells (20). The basic concept of the method consists of transforming the differential equations, which govern the physical problem, into a system of algebraic equations by applying the difference operator. This operator transforms all continuous derivatives into the ratio of changes in the variables over a small but finite element. The nodal points where unknowns are associated have to be prescribed and interpolation functions, which govern the derivatives, need to be specified. The final system of equations is constituted by both a narrow banded matrix and an independent vector which is computed according to certain prescribed boundary conditions.

Since the early sixties, when the finite element method was introduced (21, 22), it has been the most common procedure employed in geotechnical engineering. The technique consists of subdividing the domain (physical discretization) into elements and treating each of them as a smaller continuum. Nodes should be associated internal or externally to each element, and trial functions, usually polynomials, are chosen to approximate the real response of the domain. Taking into account that these functions are uniquely defined at the nodal points, the finite element equations are easily obtained by several different approaches such as the variational principle and weighted residual technique. The use of any of these procedures results in a normally banded and symmetric system of equations. The method is undoubtedly more efficient than the finite difference technique for many practical applications. In the last twenty years the method has experienced a tremendous growth in both theoretical developments and applications, and has reached such a stage that a very wide range of linear and nonlinear problems is now being solved.

The finite difference and finite element methods are known as the "domain" type technique in which discretizations are introduced in the whole domain. An alternative to reduce the type and number of these approximations is given by the so-called Boundary techniques in which the main discretizations are imposed only on the boundary of the system under consideration. These techniques consist of transforming the governing partial differential equation, which involves the behaviour of the unknown solution inside and on the surface of the domain, into an integral equation relating only boundary values. As a direct consequence, the dimension of the problem is reduced by one. The surface of the domain is discretized into elements, and polynomial functions equivalent to those used in finite elements are then necessary to interpolate the values of approximated solution between nodal points. As the unknowns of the system are related only to the boundary, the domain does not need to be discretized. After performing the boundary integrals which are done mainly by some numerical processes(23,24) smaller systems of equations are obtained compared with those achieved by domain type techniques. Then other sought responses at specified points inside the body or on the boundary can be evaluated after solving the system of equations.

The boundary integral equation method - BIEM - was one of the first techniques showing the above pattern to be applied in elastostatic problems (4). This technique still remains among the boundary methods and its theory can be seen in references (5,6). Recently the boundary element - BEM - (1,2,3) has been formulated following the same principles which govern BIEM. The method's name is due to the "element" type discretization used on the boundary. The integral equation can now also be formulated using weighted residual technique (7) which allows BEM to be related to domain type methods such as finite elements.

In recent years the boundary techniques are increasingly gaining popularity among numerical methods. The main reasons for this growth are : i) reduced set of equations; ii) a smaller amount of data; iii) proper modelling of infinite domains; iv) no interpolation error inside the domain; v) and valuable representation for stress concentration problems.

The boundary methods can be of as many different types as the domain methods, ranging from simple techniques such as the so-called indirect methods to the more versatile direct formulations.

The indirect formulations are known as the least sophisticated boundary methods. Their simplest form consists of using unit singular solutions which satisfy the governing differential equation of the problem in the domain, with specified unknown densities. Such densities have no physical significance, but once they have been obtained by enforcing boundary conditions at a number of points, the displacements and stresses can be readily computed.

Kupradze (8) established the foundations of the indirect boundary element method adopting the Kelvin fundamental solution (28) to solve elastostatic problems. Watson (9, 25, 26), applying the same technique, has obtained a numerical solution for a prestressed concrete pressure vessel of a nuclear reactor treating the structure as a thick shell.

The starting point of the direct formulation is due to Rizzo (4) in 1967. His work, which presents the solution of the two-dimensional elastostatic problem, was extended by Cruse (27) to the three-dimensional case.

The direct boundary element methods, which are more reliable than indirect techniques, are based on adopting the real physical variables of the problem as the unknown of the system. For instance, in elasticity, tractions and displacements are directly obtained by solving the system of equations. Then, other values of displacements and stresses are also computed directly by using the boundary values obtained previously.

In linear elastic application, one of the most common fundamental solutions to the corresponding differential equation of the problem is due to Lord Kelvin (28), and corresponds to a unit point load applied to the infinite domain. Other formulations adopting different fundamental solutions have already been suggested trying a better modelling for some particular problems. For this kind of formulation, usually certain boundary conditions which should be attended by the problem under consideration, are introduced in the fundamental solutions. Nakaguma (29) has presented a formulation which takes into account the free surface of the half-space problems for the direct boundary element method to solve three-dimensional problems in elastostatics. In this case, the Mindlin (30) solution for a single unit point load was used and the results obtained compared well with usual formulations in which considerably more elements were needed. A similar solution has also been presented by Telles and Brebbia (31) for the two-dimensional case, using the Melan (32) fundamental solution. Many other cases with relevant applications enforcing special conditions to the fundamental solution can be seen throughout the boundary element literature, and some of these approaches are in references (33, 56).

Only a small number of publications using boundary element techniques have been presented in conjunction with nonlinear analysis, specially in the geomechanical context. In metal applications, some works have been published since the beginning of the seventies and throughout the last decade (34, 35, 36, 37). More recently Telles and Brebbia (38, 39, 40) suggested an economical scheme to solve elastoplastic and elasto/viscoplastic problems by adapting the initial strain (41) and stress (42) processes to boundary element formulations. They have introduced the appropriated concept of singular integral derivatives based on a work carried out by Mikhlin (43).

The main objective of this work is concerned with the application of boundary element method to nonlinear problems in geomechanical stress analysis. The complete formulation for two-dimensional piecewise homogeneous analysis is presented, including plane stress, plane strain and complete plane strain cases. Several material complexities which were discussed at the beginning of this chapter are included for the analysis of rock and soil systems. Particularly for rock mechanics the no-tension material and the discontinuity analysis were proposed in conjunction with the boundary element formulation. In addition, elastoplastic and elasto/viscoplastic analyses have also been extended to solve geomechanical problems with special reference to practical bearing capacity and tunnelling applications.

CHAPTER 2

MATERIAL BEHAVIOUR AND NUMERICAL TECHNIQUES

2.1 Introduction

Many stress analyses for geomechanical materials have been successfully carried out employing numerical techniques. Since the initial growth in electronic computer science in the sixties, the finite element method has been the most common technique to analyse stress distributions and deformation patterns in tunnel excavations, embankments, dams, building foundations and other geomechanical problems.

Although the finite elements have been used in so many practical problems, the boundary formulations appear as an alternative technique which in many cases can provide more reliable or economical analysis. The excessive amount of approximation introduced by the former technique to solve problems in continuum mechanics due to the domain discretization, is avoided in boundary techniques.

The amount of information obtained from a single solution is another point in its favour. In boundary elements only the necessary information required by the analyst is computed. Displacement and stress profiles can be obtained for any part of the body, but can be avoided when the information is not necessary. In addition, the infinite domain problems which are frequently found in geomechanical stress analysis, can be fairly represented by the formulation.

In rock and soil application, the choice of a realistic material behaviour represents a great obstacle for numerical techniques.

Also the determination of material parameters to be used in the analysis can have significant influence in finite or boundary element solutions. So, many criteria to model geomechanical material response have been established so far, and often applied in conjunction with numerical techniques. We do not intend here to cover all possible choices or cases, but some significant and realistic approaches which have already been suggested and used in geomechanical stress analysis will be briefly reviewed.

Broadly speaking, the material behaviour choices which have already been implemented for numerical analysis can be divided into four main groups:-

(a)　　Linear elastic material problems.
(b)　　Nonlinear elastic material problems.
(c)　　Inelastic material problems.
(d)　　Time-dependent problems.

These four considerations upon the material behaviour and some of the important works in conjunction with numerical techniques will be reviewed in the following sections. Whenever possible, the relevant references using boundary element formulation will be pointed out in this context.

2.2 Linear Elastic Material Problems

When numerical techniques experienced initial growth in the sixties, all structural and continuum mechanics problems were treated as linear elastic cases. Even now, when other more reliable assumptions can be made, the linear theory is still necessary in many geomechanical stress analyses.

A complex work, for example tunnel or dam construction, must be carefully analysed in several stages. For many reasons, linear elastic analyses are usually adopted for some studies. For instance, data roughness is usually the reason which leads the geotechnical analyst to assume linear elastic behaviour in preliminary studies of dam design. Many other reasons can be stated to illustrate the usefulness of linear theory in geomechanical works.

Linear elastic theory has been applied in rock and soil material since the sixties, in conjunction with finite element formulation. One such analysis, which is related here as a starting point, is due to Clough (58), who investigated the stress distribution in a semi-infinite elastic half-space subjected to a concentrated load point.

The boundary element formulation in two or three-dimensional continuum mechanics was introduced in the end of the sixties (4, 27). Since then, few relevant works have been developed using boundary techniques to solve geomechanical problems.

In 1976, Hocking (59) presented a study of three-dimensional excavations based on formulations developed by Lachat and Watson (60). The author has paid special attention to the stress concentration around a flat end of a borehole for the determination of the rock stress field. He also analysed the problem of the face advancing and the overstress state which can lead the material to develop fractured zones near the excavation.

Brady and Bray (61) have presented, with an indirect formulation, a technique to solve tunnel problems in which the excavation axis is not coincident with a principal stress direction. Special considerations on the third direction derivatives were adopted in the three-dimensional governing equations for elastostatics. As a result, two uncoupled problems are obtained: normal plane strain and anti-plane problems. The final solution is given by adding the results obtained from both anti-plane and plane problems. The formulation has been extended to direct formulation (62) in 1979 and applied to tunnel analysis.

The type of singularities called quadrupoles, to take into account proximity of adjacent boundary, has been introduced in the indirect boundary formulation (64) in 1978. This modified formulation has been used to study stress distribution and lining induced displacements around a long narrow parallel-sided opening in an elastic rock medium. The same technique has been implemented by Hocking (65) in 1979 to analyse thin seam problems. As in (64), the quadrupole type of singularities was used to take into account the proximities of adjacent boundary elements due to the small thickness shown by seam inclusion in rock medium.

Wardle and Crotty (66) have solved excavation problems in rock medium using a direct formulation for two-dimensional piece-homogeneous body. The approach takes into account the traction discontinuities shown in sharp nodes on interfaces. For this purpose special considerations relating displacements and tractions of sharp nodes were introduced according to a work previously carried out by Chadonneret (67).

Nakaguma (29) in 1979 has presented a three-dimensional formulation for the direct boundary element method in which the

Mindlin (30) fundamental solution is used. In this approach, consideration of the free surface in half-space problem is already taken into account by the fundamental solution. As a result, less interpolation in the unknown displacements is introduced in the resolution of the problem and a smaller system of equations is obtained In this work, several soil structure interaction applications were carried out. The results, although obtained with constant interpolation function, show the efficiency of the technique for elastostatic problems.

The boundary element formulation has already been used in conjunction with other numerical techniques. In geomechanic stress analysis Brady and Wassing (68) have presented a two-dimensional formulation to solve tunnel problems. In three-dimensional formulation, Dendrou (69) has shown an approach to determinate effects on the liner of an underground intersection. The formulation also takes into account loosening conditions along preexisting surfaces of discontinuities, which usually result in changes in the equilibrium state.

2.3 Nonlinear Elastic Material Problem

A clear disadvantage of using linear assumption in any soil or rock problem is concerned with the reliability of the safety factor (F.S.). In a more modern concept F.S. is defined according to the limit load capacity or rupture limit. In this way F.S. gives the amount of extra load which might be applied to the system before its failure. In linear elastic analysis the safety factor, usually monitored by some chosen stress limits, cannot be a reliable parameter when associated with system failure.

Based on this consideration, the assumption of more appropriate theories to be adopted in conjunction with numerical techniques was necessary in many geomechanical stress analysis problems. Nonlinear elastic theories are among the first models used together with numerical methods in that context. The important approaches which have been developed for finite element formulations are now reviewed.

The first reported use of such a formulation was carried out in 1967 when Clough and Woodward (70) used nonlinear elasticity to study the development of stresses and deformations in the Otter Brock Dam during the construction period. The finite element method was used to model the incremental construction scheme which was represented by a sequence of layers. This incremental procedure showed the nonlinearity incorporated into the solution. The process adopted both a constant bulk modulus equal to its initial value and a variable shear modulus, which is a function of the construction scheme to model the nonlinear behaviour.

Following the initial nonlinear procedure shown above, other approaches concerned with nonlinear elastic soil and rock materials have been developed. The bi-linear and multi-linear models are among them. In the first case, the stress-strain curve of the material is modelled by two straight lines (fig. 2.3.1). The material has an initial modulus until the stresses reach a yield value, σ_y ,after which the modulus is changed. Dunlop and Duncan (71) in 1970 have studied the development of failure around excavated slopes in clay under undrained conditions using this approach. In their studies, the Tresca yield criterion was used to monitor the bilinear model, and the excavation was simulated by considering it to be executed in several stages.

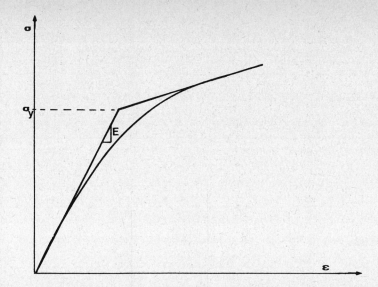

Figure 2.3.1 Bi-Linear Model.

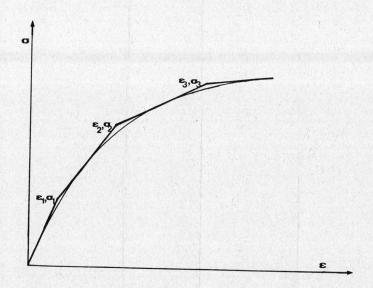

Figure 2.3.2 Multi-Linear Model.

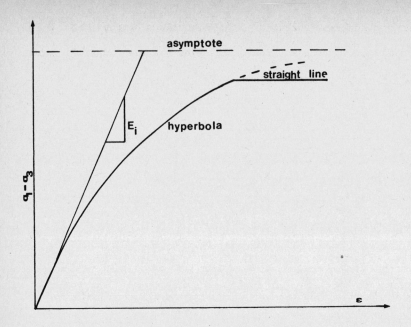

Figure 2.3.3 Hyperbolic Model.

The so-called multi-linear or piecewise linear models (72) are obtained when the stress-strain curve is given by several straight segments. In the initial stage of the use of the finite element method in geotechnical engineering, the multi-linear approach involved interpolation procedures in which the tangent moduli were computed on the basis of a set of data points $(\sigma_i, \varepsilon_i)$ on the given stress-strain curve (fig. 2.3.2).

The most used nonlinear elastic simulation of stress-strain curves in finite element analysis was developed by Duncan et al (73, 74). In their approach, the Kondner's hyperbolic model was used to describe the stress-strain relationship. Such a model is described by a hyperbolic curve for the initial section, and by a straight line for the second part of the stress-strain curve (fig. 2.3.3). The hyperbola is defined by two constants whose values have to be determinated experimentally. The formulation has been applied successfully in many practical cases and since its initial presentation many improvements have been introduced.

2.4 Inelastic Material Problems

In the previous section the discussed nonlinear criteria relate the stress directly to the strain even though they may be expressed in the form of an instantaneous modulus. For inelastic cases, the direct relation between stress and strain is no longer valid and incremental schemes are always required to model the material behaviour.

The finite element formulation was used in most of the cases to be discussed here. Some boundary element applications, not necessarily related to geomechanics, are also presented to emphasize the progress of the technique so far.

For rock materials, two inelastic approaches were proposed in 1968. In the first, Valliappan (46) proposed the no-tension criterion which takes into account the inability of rock mass to sustain tensile stresses. Due to the very nature of rock mass being fissured, it is conceivable that it cannot withstand any tension, although it can carry any compressive state of stress. The approach consists of applying an equivalent load corresponding to the tensile stresses in order to achieve a final solution without any tensile zone.

The second approach was presented by Goodman (75) to model jointed rocks. In this approach, the real faults are discretized and special elements are introduced in the analysis. New equations due to these elements are then assembled in the system of equations. An incremental and iterative process, in which the additional equations may be changed in function of joint conditions, is required to achieve the final solution. The approach has proved to be suitable to model any faults or natural planes of weakness inside the rock mass even when the joints are shortly spaced, forming a system constituted by small blocks.

Reyes and Deere (76) presented the first important numerical application in elastoplastic analysis in geomechanics, in 1966. They used the Prandtl-Reuss equations and the Drucker-Prager generalization of the Mohr-Coulomb yield surface to model elastoplastic behaviour for an unlined circular underground opening. Linear elastic behaviour was assumed until the stresses violated the yield criterion in one or more elements. At that stage, the stiffness of the plastic elements was modified. The tunnel excavation was modelled by the relief of the original stresses on the boundary, which have been taken constant elsewhere in the rock material. The final solution showed that they were in agreement with previous field observations.

As an alternative to the approach described above, Zienkiewicz, Valliappan and King (42) suggested a more efficient technique in which the repeated formation and solution of the stiffness matrix were avoided. In their approach, known as "the initial stress method", the matrix is kept constant and the excess of stress monitored by the yield surface is redistributed elastically by calculating the load corresponding to this stress.

Great efforts have been spent by geotechnical researchers in order to achieve more realistic models to represent the behaviour of geomechanical materials. Based on two theories, the Modified Cam Clay (77) and the Stress Dilatancy Theory of Rowe (78), Naylor and Zienkiewicz (79) carried out a numerical implementation to study a drained triaxial compression test on a hypothetical normally consolidated clay. The results obtained showed that the problem of large volumetric strain observed in other approaches was restrained.

The inclusion of strain-softening theories in geomechanical analysis is due to Höeg (80). He has applied an approach similar to Reyes and Deere's technique (76) to study the behaviour of circular foundations on saturated clay. The Tresca and von Mises yield surfaces were used and the results obtained were virtually the same for the two theories.

Known published works using boundary element formulations are more related with metal applications than with geomechanics. The first elastoplastic analysis in conjunction with boundary techniques is due to Ricardella (34) in 1973. He used the von Mises yield criterion for two-dimensional problems. Constant interpolation was employed to integrate plastic strains, and linear elements were used on the boundary. The problem of singular integral derivatives which arrives for the calculation of stress or strain at interior points was avoided by first integrating analytically the plastic strain over the domain and then obtaining the derivatives.

Mendelson et al (35, 81) have presented different approaches to model elastoplastic responses with boundary techniques. In those references, some elastoplastic results were presented such as numerical solutions for both an edge-notched beam and a square cross section bar under pure bending and torsion respectively.

Another elastoplastic approach using boundary element technique is due to Mukherjee (36). He produced modified versions for the kernels of the plastic strain integral. However, such modification introduced limitations in the approach, which can now only analyse systems where incompressibility of the plastic strain is assumed.

Telles and Brebbia (38), based on the appropriate concept for the derivative of the singular integral of the inelastic term, suggested approaches using initial strain and stress processes to model plastic behaviour with the boundary element method. Their first approach follows the Mendelson's process (41) in which an "initial strain" form of the inelastic term is considered. This formulation is only capable of handling incompressible plastic strain using the isotropic von Mises yield criterion.

In reference (39) the authors presented an "initial stress" approach using four different yield criteria (Tresca, von Mises, Mohr-Coulomb and Drucker-Prager). In this approach the first results of an elastoplastic stress analysis were shown and they correspond to the solution of the same underground opening used in Reyes and Deere's work (76). The obtained results clearly demonstrate the suitability of the boundary element method for elastoplastic stress analysis in an infinite rock medium.

The analysis of slip and separation along discontinuities was treated by Hocking (65). The author has considered this nonlinear effect dealing with the same equations which have been proposed to solve the seam case in the same work by introducing some particularizations to model sliding or separation along joints.

The no-tension approach already described for finite elements has also been extended to boundary element formulations (82, 83, 84). In this approach, elastic and inelastic considerations have been made and good agreement with analytical and finite element solutions has been verified.

2.5 Time-Dependent Problems

Most of rock and soil materials exhibit a delayed deformation pattern when subjected to any load. Many researches have already been carried out trying to find realistic time-dependent behaviours for such materials. In particular, Lama and Vutukuri (14) have given extensive overviews of many time-dependent theories formulated to model rheological behaviour of rocks. Viscoelastic responses which can be associated with Maxwell or Kelvin models (fig. 2.5.1) have been adopted several times to study long-time deformations in geomechanical problems using numerical techniques. One of the first applications of such an approach in rock material design was carried out by Zienkiewicz, Watson and King (85) using a finite element formulation. In this approach the material behaviour is adequately represented by a finite series of Kelvin elements coupled with elastic responses. The elastic strain pattern and the solution for stress distribution were obtained applying an initial strain field due to the viscous effects to the system. In this publication, they presented a viscoelastic analysis of a lined tunnel constructed in a rock medium.

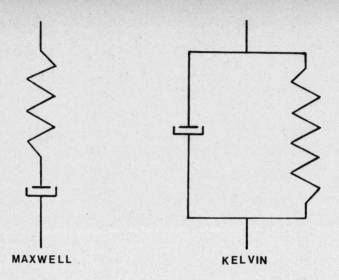

Figure 2.5.1 Maxwell and Kelvin Models.

Several other representations for time-dependent problems have already been suggested in the literature, most of them obtained by adopting different combinations of Kelvin, Maxwell and Bingham's elements. One such combination consists of two Bingham's units associated in series and was suggested by Lombardi (86). The author has employed his formulation to simulate rock movements due to a tunnel excavation. He claimed that the model has successfully represented the creep and the failure behaviours and the results were in total agreement with field measurements taken during and after the tunnel construction.

An improvement in the viscous behaviour in numerical applications is due to Cormeau (18). In his work, he has given a complete explanation on phenomenological theories with and without postulated yield surfaces. In particular, the Perzyna's model has been adopted in his formulation to model viscoplastic, pure creep and pure relaxation responses. Geomechanical analysis based on this theory has also been carried out in the same work and in a more particularized way in reference (17). In this work, a tunnel construction has been modelled by considering the actual viscous effect between each stage of the construction, although material parameters have not been taken from a real case.

Very few formulations of the boundary element techniques to model viscous response have been published so far. It appears to be Chadonneret (37) who first used a direct boundary formulation for viscoplastic analysis. In her study, original constitutive equations, developed at O.N.E.R.A. (France) were employed to study a notched-plate and the result obtained was in agreement with the corresponding experimental analysis. The integral equations presented are based on an "initial stress" form of the viscoplastic strain influence and the numerical implementation was carried out using linear boundary elements and constant rectangular cells.

It is due to Telles and Brebbia (40) the implementation of the Perzyna's model to boundary element formulations. The solution routine employed a simple Euler time integration scheme to calculate the viscoplastic stress increment which is used to correct the total stress field.

Venturini and Brebbia (87) have extended the approach above to model more complex material responses, adopting the overlay technique (47, 48, 49) to the boundary element formulation. In their approach, the same tunnel construction simulation presented in reference (17) was analysed. The authors have also included an alternative representation for the viscous effects using overlay models.

CHAPTER 3

BOUNDARY INTEGRAL EQUATIONS

3.1 Introduction

This chapter is concerned with the introduction of the basic integral equations for two-dimensional elastic linear material problems. It starts by briefly reviewing the partial differential equations for linear elastic material and introducing the necessary notations involved in the formulation. These governing equations are also extended to deal with problems in which initial stress and strain type loads are applied. Such kind of loads are not only important to take into account temperature or other similar loads, but also to model nonlinear material behaviour when used in conjunction with a well established successive elastic solution technique.

The basic integral equations for the displacement and stress representations in an elastic body governed by the Navier equation with certain prescribed boundary conditions are presented followed by the technique to deal with problems where the body and prestress forces are applied.

3.2 Governing equations and fundamental solutions

Before discussing the governing equations for an elastic body, the necessary notation to be used here and throughout this work is introduced. The so-called Cartesian tensor notation will be used to avoid writing long expressions whenever possible. This representation is extremely useful for theorem proofs and usually long sets of equations can be represented by one single expression. The notation works with subscript or superscript indices to represent the conventional Cartesian system of coordinates (x,y,z). For an easier undertaking of the expressions involved in this work, the following rules are adopted:-

(a)   Summation Convention

Repetition of an index (whether superscript or subscript) in a term will denote a summation with respect to that index over its range, for example,

$$a^i b^i = a^1 b^1 + a^2 b^2 + a^3 b^3 \quad (i=1,2,3) \qquad (3.2.1)$$

(b)   Partial Differentiation Convention

The partial derivative of a variable with respect to a Cartesian coordinate is denoted by a comma. Thus,

$$\frac{\partial f_i}{\partial x_j} = f_{i,j} \; , \; \frac{\partial \phi}{\partial x_i} = \phi_{,i} \quad \text{and} \quad \frac{\partial \sigma_{ij}}{\partial x_k} = \sigma_{ij,k} \qquad (3.2.2)$$

(c)   Kronecker Delta δ_{ij}

This is an auxiliary parameter which assumes the following values,

$$\delta_{ij} = 0 \quad \text{if} \quad i \neq j$$
$$\delta_{ij} = 1 \quad \text{if} \quad i = j \qquad (3.2.3)$$

In this section all indices should be assumed to have a range of two or three according to the Euclidean space adopted for the analysis.

When different ranges are needed, they are locally stated.

The following elastic constants are used in this context:-

(a) Modulus of elasticity or Young's modulus : E
(b) Poisson's ratio : ν
(c) Modulus of elasticity in shear, shear modulus or modulus of rigidity : G

For the definition of the elastic problem one needs only two of the above constants. Thus, a relationship between them should be defined.

$$G = \frac{E}{2(1+\nu)} \qquad (3.2.4)$$

All the values given above are used for three-dimensional analysis and for the plane strain case. When plane stress condition is assumed for the two-dimensional analysis, equivalent values have to be defined to maintain the same formulation for both plane strain and plane stress cases. Thus,

$$\bar{G} = G$$

$$\bar{E} = E(1+2\nu)/(1+\nu)^2 \qquad (3.2.5)$$

$$\bar{\nu} = \nu/(1+\nu)$$

Considering an elastic body defined by a domain Ω and a boundary Γ, its partial equilibrium is established by analysing a system of forces and momenta for a generic point. Thus,

$$\sigma_{ij,j} + b_i = 0 \qquad (3.2.6)$$

where σ_{ij} are the components of the stress tensor and b_i stands for body force values which are defined as forces acting on elements of volume or mass inside the body.

In this particular context, the symmetry condition of the stress tensor is assumed valid. Thus,

$$\sigma_{ij} = \sigma_{ji} \qquad (3.2.7)$$

which therefore reduces the nine stress tensor components into only six independent values.

The traction vector components which are required in the following section can be defined by the relation,

$$p_i = \sigma_{ij} n_j \qquad (3.2.8)$$

where n_j are the direction cosines of the outward normal to the tangent plane at the boundary point under consideration.

Under the action of a system of forces the body Ω is displaced from its original state. Denoting u_i the displacement components within the context of small strain theory, the following strain-displacement relationship can be written,

$$\varepsilon_{ij} = \frac{1}{2}(u_{i,j} + u_{j,i}) \qquad (3.2.9)$$

The strain tensor ε_{ij} can be interpreted as being computed by two different parts,

$$\varepsilon_{ij} = \varepsilon_{ij}^e + \varepsilon_{ij}^o \qquad (3.2.10)$$

where the first value ε_{ij}^e represents the elastic deformations which occur under action of body forces or other boundary prescribed conditions.

The second term, ε_{ij}^o, usually defined as "initial strain values", represents the deformation due to the action of deformation type load, such as temperature, shrinkage, swelling, etc. The above defined initial strain term is also useful when applied together with a successive elastic solution technique to model nonlinear material behaviour.

The relationship between stress and strain tensor components for an isotropic body subject to any load is given by the general Hooke's law which is expressed by,

$$\sigma_{ij} = \frac{2G\nu}{(1-2\nu)} (\varepsilon_{kk} - \varepsilon_{\ell\ell}^o) \delta_{ij} + 2G(\varepsilon_{ij} - \varepsilon_{ij}^o) \qquad (3.2.11)$$

in which the range of the subscript "ℓ" for two-dimensional cases and plane strain conditions is $\ell = 1,2,3$. For instance, only the total deformation at the third direction is zero. The elastic (ε_{ij}^e) and the initial value (ε_{ij}^o) components can assume any values.

Alternatively, equation (3.2.11) can be written as,

$$\sigma_{ij} = C_{ijk\ell} (\varepsilon_{k\ell} - \varepsilon_{k\ell}^o) \qquad (3.2.12)$$

in which $C_{ijk\ell}$ are the elastic compliances given by,

$$C_{ijk\ell} = \frac{2G\nu}{1-2\nu} \delta_{ij} \delta_{k\ell} + G(\delta_{ik} \delta_{j\ell} + \delta_{i\ell} \delta_{jk}) \qquad (3.2.13)$$

Using expression (3.2.9), Hooke's law (3.2.11) is now written in terms of displacements,

$$\sigma_{ij} = \frac{2G\nu}{1-2\nu} u_{k,k} \delta_{ij} + G(u_{i,j} + u_{j,i}) - \frac{2G\nu}{1-2\nu} \varepsilon_{\ell\ell}^o \delta_{ij} - 2G\varepsilon_{ij}^o \qquad (3.2.14)$$

And similarly the traction vector components (eq. 3.2.8) can also be rewritten as,

$$p_i = \frac{2G\nu}{1-2\nu} u_{k,k}\eta_i + G(u_{j,i}\eta_j + u_{i,n}) - \frac{2G\nu}{1-2\nu}\varepsilon^o_{\ell\ell}\eta_i - 2G\varepsilon^o_{ij}\eta_j \quad (3.2.15)$$

where

$$u_{i,n} = u_{i,j}\eta_j \quad (3.2.16)$$

The deformation type load effects can also be modelled using an equivalent initial stress term (σ^o_{ij}). Then equation (3.2.14) becomes,

$$\sigma_{ij} = \frac{2G\nu}{1-2\nu} u_{k,k}\delta_{ij} + G(u_{i,j}+u_{j,i}) - \sigma^o_{ij} \quad (3.2.17)$$

In cases where the deformation fields are imposed, for example the temperature load, the initial stress field is obtained by,

$$\sigma^o_{ij} = \frac{2G\nu}{1-2\nu}\varepsilon^o_{\ell\ell}\delta_{ij} - 2G\varepsilon^o_{ij} \quad (3.2.18)$$

or as it occurs in some nonlinear applications obtained by applying a successive elastic solution technique, the initial stress values can be computed directly by their constitutive laws.

By introducing equation (3.2.17) into equilibrium equation (3.2.6) one obtains,

$$\frac{1}{1-2\nu} u_{j,ij} + u_{i,jj} + \frac{b_i}{G} - \frac{\sigma^o_{ij,j}}{G} = 0 \quad (3.2.19)$$

which is the Navier equation in terms of displacements. The independent term can now be interpreted as equivalent body force values as follows,

$$\bar{b}_i = b_i - \sigma^o_{ij,j} \quad (3.2.20)$$

Expression (3.2.15) can also be written including the initial stress term,

$$p_i + \sigma^o_{ij} n_j = \frac{2G\nu}{1-2\nu} u_{k,k} n_i + G(u_{j,i} n_j + u_{i,n}) \qquad (3.2.21)$$

in which the left hand side can be interpreted as an equivalent traction vector $\bar{p}_i$, i.e.,

$$\bar{p}_i = p_i + \sigma^o_{ij} n_j \qquad (3.3.33)$$

To complete the definition of the elastic body under consideration, boundary conditions must be specified. Thus,

$$u_i(Q) = \bar{u}_i(Q), \quad Q \in \Gamma_1$$
$$p_i(S) = \bar{p}_i(S), \quad S \in \Gamma_2 \qquad (3.2.23)$$

where $\Gamma = \Gamma_1 + \Gamma_2$, and $\bar{u}_i$ and $\bar{p}_i$ are prescribed components of the displacements and tractions on the whole boundary Γ.

In this context, the so-called fundamental solutions will be extensively used from now on. Although several different solutions can be applied in the formulation, here only one due to Lord Kelvin (28) is used. To obtain this solution let us consider an infinite space which is represented by a domain Ω^* and a boundary Γ^*. The solution which is being sought is defined as the responses at point "q" (field point) due to a unit point load applied at "s" (load point) in the direction of the Cartesian axes, with Ω^* containing the points "q" and "s" (fig. 3.2.1). The problem is formulated by using the Navier equation (3.2.19) with the body force terms replaced by the Dirac delta function

$\delta(s,q)$. Thus,

$$\bar{b}_i(q) = \delta(s,q)c_i(s) \tag{3.2.24}$$

where $c_i(s)$ are a set of unit vectors in the x_i direction; and also

$$\delta(s,q) = 0 \quad \text{if} \quad q \neq s$$
$$\delta(s,q) = \infty \quad \text{if} \quad q = s \tag{3.2.25}$$

$$\int_{\Omega^*} g(q)\delta(s,q)d\Omega^*(q) = g(s)$$

Therefore, the equilibrium equation (3.2.6) can be written for that particular case as follows,

$$\sigma^*_{ij,j} + \delta(s,q)\,c_i(s) = 0 \tag{3.2.26}$$

An analogous expression can also be formulated in terms of displacements if equation (3.2.19) were used. Then,

$$\frac{1}{1-2\nu} u^*_{j,ij} + u^*_{i,jj} + \frac{1}{G}\delta(s,q)\,c_i(s) = 0 \tag{3.2.27}$$

in which the symbol "*" is used to indicate the fundamental problem.

The appropriate expression for the displacements, u^*_{ij}, which obey equation (3.2.27) is the known Kelvin fundamental solution and is presented in its tensor form (1,2) as follows,

$$u^*_{ij}(s,q) = \frac{1}{16\pi(1-\nu)G}\left[(3-4\nu)\delta_{ij} + r_{,i}r_{,j}\right] \tag{3.2.28}$$

for three-dimensional problems, and

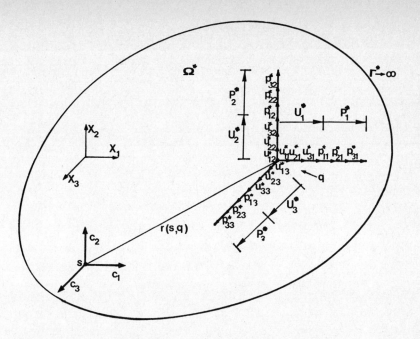

Figure 3.2.1 Fundamental Solutions.

$$u^*_{ij}(s,q) = \frac{-1}{8\pi(1-\nu)G}\left[(3-4\nu)\ln r \delta_{ij} - r,_i r,_j\right] \quad (3.2.29)$$

for plane strain problems, in which the first index corresponds to the direction of the point load and the second to the component of the respective displacement (fig. 3.2.1).

Equation (3.2.21) can also express the fundamental solution for plane stress problems, since the Poisson's ratio ν is replaced by its equivalent value $\bar{\nu}$ given in equation (3.2.5).

The fundamental solution for traction can be evaluated for a given surface through equation (3.2.21), and its final expression becomes,

$$p_{ij}(s,q) = \frac{-1}{4\alpha\pi(1-\nu)r^\alpha}\left\{\left[(1-2\nu)\delta_{ij} + \beta r,_i r,_j\right]\frac{\partial r}{\partial n}\right. \quad (3.2.30)$$

$$\left. - (1-2\nu)(r,_i n_j - r,_j n_i)\right\}$$

from where two and three-dimensional solutions are obtained letting $\alpha = 1,2$ and $\beta = 2,3$ respectively. Furthermore, the derivatives $r_{,i}$ are taken with reference to the coordinates of the field point "q". Thus,

$$r_{,i} = \frac{\partial r}{\partial x_i(q)} = \frac{x_i(q)-x_i(s)}{r} \qquad (3.2.31)$$

and

$$\frac{\partial r}{\partial n} = r_{,i} n_i \qquad (3.2.32)$$

3.3 Integral Equations

In this section, integral equations for displacement and stress determinations at any point a two-dimensional elastic body are presented. The displacement for an interior point "s" know as the somigliana's identity (44) can be obtained using either an extended version of the Betti's second work theorem (45) or the weighted residual technique (7). Based on one of these concepts the following expression is easily derived (1,2),

$$u_i(s) = -\int_\Gamma u_k(Q) p^*_{ik}(s,Q) d\Gamma(Q) + \int_\Gamma p_k(Q) u^*_{ik}(s,Q) d\Gamma(Q) \qquad (3.3.1)$$

$$+ \int_\Omega b_k(q) u^*_{ik}(s,q) d\Omega(q) + \int_\Omega \sigma^o_{mk}(q) \varepsilon^*_{imk}(s,q) d\Gamma(q)$$

where "s" and "q" ε Ω (domain), Q ε Γ (boundary), u_k and p_k are the displacement and traction values over the boundary, b_k and σ^o_{mk} are the body force and initial stress components, and ε^*_{imk} (s.9) is given by,

$$\varepsilon^*_{imk}(s,q) = \frac{-1}{8\pi(1-\nu)Gr} \left\{ (1-2\nu)(r,_k \delta_{im} + r,_m \delta_{ik}) - r,_i \delta_{mk} + 2r,_i r,_m r,_k \right\} \quad (3.2)$$

for plane strain condition. For plane stress, one has to use $\bar{\nu}$ instead of ν.

The Somigliana's identity for displacements presented above takes into consideration an initial stress field applied in the body. Similar expressions can be obtained when an initial strain field is considered, i.e.,

$$u_i(s) = -\int_\Gamma u_k(Q) p^*_{ik}(s,Q) d\Gamma(Q) + \int_\Gamma p_k(Q) u^*_{ik}(s,Q) d\Gamma(Q) \quad (3.3)$$

$$+ \int_\Omega b_k(q) u^*_{ik}(s,q) d\Omega(q) + \int_\Omega \varepsilon^o_{m\ell}(q) \sigma^*_{im\ell}(s,q) d\Omega(q)$$

where the summation subscripts "m" and "ℓ" for the last integral have a range of (1,2,3) or (1,2) depending on whether plane strain or plane stress conditions are assumed, and the tensor $\sigma^*_{im\ell}(s,q)$ is obtained directly from $\varepsilon^*_{im\ell}$ by applying Hooke's law, i.e.,

$$\sigma^*_{im\ell}(s,q) = \frac{-1}{4\pi(1-\nu)r} \left\{ (1-2\nu)(r,_\ell \delta_{im} + r,_m \delta_{i\ell} - r,_i \delta_{m\ell}) + 2r,_i r,_m r,_\ell \right\} \quad (3.4)$$

The extension of these expressions to a boundary point "S" is easily obtained by increasing the domain around the point under consideration (see figure 3.3.1) in order to apply equation (3.3.1) or (3.3.3), and then taking the limit when $\varepsilon \to 0$.

The final expression for the boundary point "S" derived using this procedure is given by (1),

$$c_{ik}(S)u_k(S) + \int_\Gamma p^*_{ik}(S,Q)u_k(Q)d\Gamma(Q) = \int_\Gamma u^*_{ik}(S,Q)p_k(Q)d\Gamma(Q) \quad (3.3.5)$$

$$+ \int_\Omega u^*_{ik}(S,q)b_k(q)d(q) + \int_\Omega \varepsilon^*_{imk}(S,q)\sigma^o_{mk}(q)d(q)$$

where the coefficient $c_{ik}(S)$ is dependent only on the geometry of the boundary and is obtained by,

$$c_{ik}(S) = \delta_{ik} + \lim_{\varepsilon \to 0}\int_{\Gamma_\varepsilon} p^*_{ik}(S,Q)d\Gamma(Q) \quad (3.3.6)$$

Similar equation can be derived in the case of an initial strain field applied in the domain. In this case equation (3.3.5) is still valid since the last integral is changed as in equation (3.3.3).

Equation (3.3.5) which expresses a relation for tractions and displacements is usually derived for bounded bodies, but can be easily extended to infinite bodies (see reference 60).

Equations (3.3.1) and (3.3.3) are continuous representations for displacements at any interior point "s". Thus the state of stress can be calculated by differentiating these equations and substituting the differentiated values into equations (3.2.14) and (3.2.17).

After carrying out the steps mentioned above using the proper conpept of singular intesral derivative (43) as has been done in reference (38,90), the following equation for stress determination at an interior point is derived,

$$\sigma_{ij}(s) = -\int_\Gamma S_{ijk}(s,Q)u_k(Q)d\Gamma(Q) + \int_\Gamma D_{ijk}(s,Q)p_k(Q)d\Gamma(Q) +$$

$$\int_\Omega D_{ijk}(s,q)b_k(q)d\Omega(q) + \int_\Omega E_{ijmk}(s,q)\sigma^o_{mk}(q)d\Omega(q) \quad (3.3.7)$$

$$- \frac{1}{8(1-\nu)}\left[2\sigma^o_{ij}(s) + (1-4\nu)\sigma^o_{\ell\ell}(s)\delta_{ij}\right]$$

where the integral of the initial stress term is in the Chauchy principal value sense, and the tensors $S_{ijk}(s,Q)$, $D_{ijk}(s,Q)$ or $D_{ijk}(s,q)$, and $E_{ijmk}(s,q)$ are given below,

$$S_{ijk}(s,Q) = \frac{2G}{4\pi(1-\nu)r^2}\left\{2r_{,n}\left[(1-2\nu)\delta_{ij}r_{,k}+\nu(\delta_{ik}r_{,j}+\delta_{jk}r_{,i})-4r_{,i}r_{,j}r_{,k}\right]\right.$$

$$+ 2\nu(n_i r_{,j} r_{,k}+n_j r_{,i} r_{,k})+(1-2\nu)(2n_k r_{,i} r_{,j} + \quad (3.3.8)$$

$$\left. + n_j\delta_{ik}+n_i\delta_{jk}) - (1-4\nu)n_k\delta_{ij}\right\}$$

$$D_{ijk}(s,Q) = \frac{1}{4\pi(1-\nu)r}\left\{(1-2\nu)\left[\delta_{ik}r_{,j}+\delta_{jk}r_{,i}-\delta_{ij}r_{,k}\right]+ 2r_{,i}r_{,j}r_{,k}\right\} \quad (3.3.9)$$

$$E_{ijmk}(s,q) = \frac{1}{4\pi(1-\nu)r^2}\left\{(1-2\nu)\left[\delta_{ik}\delta_{jm}+\delta_{jk}\delta_{im}-\delta_{ij}\delta_{mk}+2\delta_{ij}r_{,m}r_{,k}\right]\right.$$

$$+ 2\nu\left[\delta_{im}r_{,j}r_{,k}+\delta_{jk}r_{,i}r_{,m}+\delta_{ik}r_{,j}r_{,m}+\delta_{jm}r_{,i}r_{,k}\right] \quad (3.3.10)$$

$$\left. + 2\delta_{mk}r_{,i}r_{,j}-8r_{,i}r_{,j}r_{,m}r_{,k}\right\}$$

Similar formulation is obtained when an initial strain field is applied to the domain under consideration. However, in this case different equations are achieved according to whether plane strain or stress conditions are assumed. For the plane strain conditions, one has to take into consideration the work performed in the third direction $[\sigma^*_{i33}(s,q)\,\varepsilon^o_{33}(q)]$ which has been indicated by the range of the subscript index in equation (3.3.3). Thus, using expressions (3.3.3) and (3.2.14) and assuming plane strain conditions, the stress values for an interior point "s" are given by,

$$\sigma_{ij}(s) = -\int_\Gamma S_{ijk}(s,Q)u_k(Q)d\Gamma(Q) + \int_\Gamma D_{ijk}(s,Q)p_k(Q)d\Gamma(Q) +$$

$$+ \int_\Omega D_{ijk}(s,q)b_k(q)d\Omega(q) + \int_\Omega F_{ijmk}(s,q)\varepsilon^o_{mk}(q)d\Omega(q) \qquad (3.3.11)$$

$$+ \int_\Omega F_{ij33}(s,q)\varepsilon^o_{33}(q)d\Omega(q) - \frac{G}{4(1-\nu)}\left[2\varepsilon^o_{ij}(s)+[\varepsilon^o_{\ell\ell}(s)+4\nu\varepsilon^o_{33}(s)]\delta_{ij}\right]$$

where $\varepsilon^o_{33}(s)$ is the third direction initial strain value; all subscript indices have a range of 2; and the tensors $F_{ijmk}(s,q)$ and $F_{ij33}(s,q)$ are given by the following expressions,

$$F_{ijmk}(s,q) = \frac{G}{2\pi(1-\nu)r^2}\Big\{ 2(1-\nu)(\delta_{ij}r,_m r,_k + \delta_{mk}r,_i r,_j) + 2\nu(\delta_{im}r,_j r,_k +$$

$$\delta_{jk}r,_i r,_m + \delta_{ik}r,_j r,_m + \delta_{jm}r,_i r,_k) - 8r,_i r,_j r,_m r,_k +$$

$$(1-2\nu)(\delta_{ik}\delta_{jm}+\delta_{jk}\delta_{im}) - (1-4\nu)\delta_{ij}\delta_{mk} \Big\} \qquad (3.3.12)$$

and

$$F_{ij33}(s,q) = \frac{-\nu G}{(1-\nu)r^2}\Big[-\delta_{ij}+2r,_i r,_j\Big] \qquad (3.3.13)$$

For plane stress conditions the same formulation is valid since ν is replaced by $\bar{\nu}$, and $F_{ij33}(s,q)$ and $\varepsilon^o_{33}(s)$ are neglected.

3.4 Body Force Problem

By examining equations (3.3.5) and (3.3.7) one can see that the body force effects are computed by means of particular integrals over the relevant domain. These integrals are usually computed by numerical processes which often very much increase the total amount of data required to solve a problem, and are also inconvenient from the computing point of view. However, as can be seen in references (93,94,95) the domain integrals can be transformed into boundary integrals which are more appropriate for boundary formulations.

Let us now examine separately the body force integral terms in both displacement and stress equations (3.3.5) and (3.3.7). Thus, for displacement determination one can write,

$$B_i(S) = \int_\Omega u^*_{ik}(S,q) b_k(q) d\Omega(q) \qquad (3.4.1)$$

and for stress,

$$\bar{B}_{ij}(s) = \int_\Omega D_{ijk}(s,q) b_k(q) d\Omega(q) \qquad (3.4.2)$$

The usual way to transform these domain integrals into boundary integrals requires the use of a special tensor known as the Galerkian tensor which modifies the displacement kernel $u^*_{ik}(s,q)$. When such a tensor is introduced into equation (3.4.1) easier arrangements can be made replacing the domain integrals by boundary integrals (95).

For many practical applications in geomechanics, the only body force load which occurs is the self-weight. Then, to take into account such a kind of load, only the constant body force case has to be analysed.

That formulation can be easily obtained by direct integration over one dimension of the domain. Thus, let one consider the domain Ω bounded by a surface Γ as indicated in fig. (3.4.1), where a constant body force field b_k is acting. Writing the body force integral (eq. 3.4.1) in terms of cylindrical coordinates, one has,

$$B_i(S) = b_k \int_\theta \int_r u^*_{ik}(S,q) r \, dr \, d\theta \qquad (3.4.3)$$

Assuming the Kelvin fundamental solution, the integral over r can be performed as follows,

$$B_i(S) = \frac{b_k}{16\pi G(1-\nu)} \int_\theta \left[(-3+4\nu)(\ln R - \frac{1}{2}) R \delta_{ik} + R,_i R,_k R \right] d\theta \qquad (3.4.4)$$

in which $R = R(S,Q)$ is a particular value of $r(S,q)$ when the point "Q" is on the boundary as indicated in fig. (3.4.1).

Taking into consideration that θ is a function of the boundary Γ, the integral (eq. 3.4.4) becomes,

$$B_i(S) = b_k \int_\Gamma B^*_{ik}(S,Q) d\Gamma \qquad (3.4.5)$$

in which

$$B^*_{ik}(S,Q) = \frac{R}{16\pi G(1-\nu)} \left[-(3-4\nu)(\ln R - \frac{1}{2}) \delta_{ik} + R,_i R,_k \right] \eta_\ell R,_\ell \qquad (3.4.6)$$

By using a similar procedure the integral term (eq. 3.4.2) for the stress determination becomes,

$$\bar{B}_{ij}(s,Q) = b_k \int_\Gamma T_{ijk}(s,Q) d\Gamma(Q) \qquad (3.4.7)$$

in which

$$T_{ijk}(s,Q) = D_{ijk}(s,Q)\eta_\ell R,_\ell \qquad (3.4.8)$$

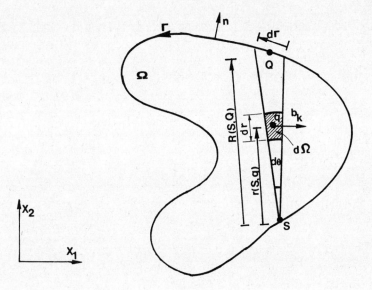

Figure 3.4.1 Finite Body. Scheme for Domain Integration.

3.5 Prestress Force Problem

The use of prestress forces has been regularly adopted in many rock analyses. This practice is usually needed either to control plastic and loosening regions around tunnels, which is achieved with the application of radial bolting in the rock mass, or to secure the stability of rock slopes.

The prestress effects are obtained by applying a system of compressive forces without resultant. In order to take into account such a kind of load in two-dimensional boundary formulation, some approximations have to be introduced to compute displacements and stresses in the domain. One of these approximations is constituted by assuming plane conditions (whether plane strain or plane stress). For instance, let us examine the case of a long opening in which prestress forces have been radially applied. As the loads act at discrete points regularly spaced in the direction of the opening, the validity and accuracy of the analysis is a direct function of the spacing adopted.

The prestress forces can be taken into account in the boundary integral formulation as tractions prescribed on discrete segments defined on domain surfaces (fig. 3.5.1), which are the normal boundary of the system on interfaces between subregions (see chapter 5). However, this procedure can generate systems with an excessive number of boundaries which may result in an inconvenient numerical analysis.

An alternative is to consider the prestress loads as body forces applied in small and discrete regions, which may be defined by the size of the rock anchors (fig. 3.5.2). Thus, writing separately the body force integral terms of equations (3.3.1 and 3.3.5) the prestress effects are given by,

$$V_i(S) = \int_{\Omega_j} u^*_{ik}(S,q) b_{v_k}(q) d\Omega_j(q) \qquad (j = 1,N) \qquad (3.5.1)$$

and analogously the corresponding term in equation (3.3.7) becomes,

$$\bar{V}_{ij}(s) = \int_{\Omega_\ell} D_{ijk}(s,q) b_{v_k}(q) d\Omega_\ell(q) \qquad (\ell = 1,N) \qquad (3.5.2)$$

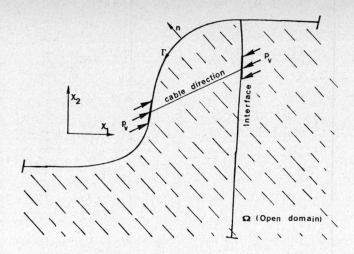

Figure 3.5.1 Prestress Forces on the Boundary.

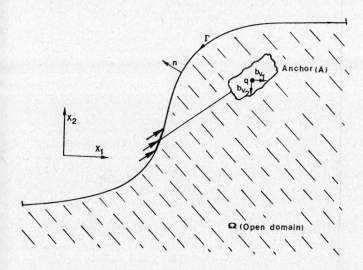

Figure 3.5.2 Prestress Forces as Body Forces Over the Anchor Area.

where N is the number of applied prestress forces, and $b_{v_k}(q)$ can be a constant value which is obtained by dividing the prestress force values P_{v_k} by the area of the anchor, i.e.,

$$b_{v_k}(q) = b_{v_k} = P_{v_k}/A \qquad (3.5.3)$$

That formulation using integrals over discrete areas of the domain can be avoided if the prestress forces were assumed to be applied at discrete points. For this case the final expression can be obtained either by performing the limit of the integrals in equations (3.5.1) and (3.5.2) when $A \to 0$, or by using the proper concept of fundamental solution. Thus, one can write,

$$V_i(S) = u^*_{ik}(S,\ell) p^\ell_{v_k} \qquad (\ell = 1,N) \qquad (3.5.4)$$

and

$$\bar{V}_{ij}(s) = D_{ijk}(s,\ell) p^\ell_{v_k} \qquad (\ell = 1,N) \qquad (3.5.5)$$

in which ℓ defines the point where the prestress forces are applied.

As the forces now are assumed to be concentrated, values of displacements and stresses are no longer possible at prestress load points. If the stresses or displacements at these points were required, approximations involving the real size of the anchor have to be made.

3.6 Temperature, Shrinkage and Swelling

In section (3.3) expressions for the determination of stresses and displacements were formulated. In these formulations the effects of an initial stress or strain field are computed by domain integrals which can be performed analytically or numerically, since nodal values are known at some discrete points or given by any other representation. For many practical applications these kinds of loading can be represented by known analytical functions; when they are dependent only on the coordinate system, the transformation of the domain integrals into boundary integrals can be obtained. Herein, the particular case of constant initial stress values is analysed. Such a representation is a usual and justified approximation, especially for modelling shrinkage effects. In this case, the transformation of the domain integral due to the initial stress field in equation (3.3.5) is performed by using a generalized form of the Gauss' theorem, which may be expressed as follows,

$$\int_\Omega F_{jk\ell,i}(s,q) d\Omega(q) = \int_\Gamma F_{jk\ell}(s,Q) \eta_i(Q) d\Gamma(Q) \qquad (3.6.1)$$

Let us therefore examine separately the last domain integral in equation (3.3.5), taking into account constant values for $\sigma^o_{mk}(q)$. Thus,

$$S_i(S) = \sigma^o_{mk} \int_\Omega \varepsilon^*_{imk}(S,q) d\Omega(q) \qquad (3.6.2)$$

Bearing in mind the definition of $\varepsilon^*_{imk}(S,q)$ (eq. 3.2.9) and applying equation (3.6.1) one obtains,

$$S_i(S) = \frac{1}{2} \sigma^o_{mk} \int_\Gamma \left[u^*_{im}(S,Q)\eta_k(Q) + u^*_{ik}(S,Q)\eta_m(Q) \right] d\Gamma(Q) \qquad (3.6.3)$$

which, after the substitution of the fundamental solution given by equation (3.2.29) becomes,

$$S_i(S) = \sigma^o_{mk} \int_\Gamma e^*_{imk}(S,Q) d\Gamma(Q) \qquad (3.6.4)$$

with

$$e^*_{imk}(S,Q) = \frac{1}{16\pi G(1-\nu)} \left\{ -(3-4\nu)\ln r \left[\delta_{im}\eta_k + \delta_{ik}\eta_m\right] + r_{,i}\left[r_{,k}\eta_m + r_{,m}\eta_k\right]\right\} \quad (3.6.5)$$

For the stress one needs only to find the partial derivatives of the above equation at an interior point "s",

$$\frac{\partial S_i(s)}{\partial x_j(s)} = \sigma^o_{mk} \int_\Gamma \frac{\partial}{\partial x_j(s)} \left[e^*_{imk}(s,Q)\right] d\Gamma(Q) \quad (3.6.6)$$

and then apply Hookes's law,

$$\bar{S}_{ij}(s) = \frac{2G\nu}{1-2\nu} \frac{\partial S_\ell(s)}{\partial x_\ell(s)} + G\left\{\frac{\partial S_i(s)}{\partial x_j(s)} + \frac{\partial S_j(s)}{\partial x_i(s)}\right\} \quad (3.6.7)$$

which, after carrying out the derivatives, becomes

$$\bar{S}_{ij}(s) = \sigma^o_{mk} \int_\Gamma f_{ijmk}(s,Q) d\Gamma(Q) \quad (3.6.8)$$

with

$$f_{ijmk}(s,Q) = \frac{1}{8\pi(1-\nu)r} \left\{ (1-2\nu)\left[r_{,i}(\delta_{jm}\eta_k + \delta_{jk}\eta_m) + r_{,j}(\delta_{im}\eta_k + \delta_{ik}\eta_m) \right.\right.$$

$$\left.\left. -\delta_{ij}(r_{,m}\eta_k + r_{,k}\eta_m)\right] + 2r_{,i}r_{,j}(r_{,k}\eta_m + r_{,m}\eta_k)\right\} \quad (3.6.9)$$

A similar formulation can be written for the initial strain case. However, the additional integral due to the deformation at the third direction has also to be considered.

CHAPTER 4

BOUNDARY INTEGRAL EQUATIONS FOR COMPLETE PLANE STRAIN PROBLEMS

4.1 Introduction

In this chapter the general case of plane strain condition known as the "complete plane strain" (61, 62) is introduced into the boundary formulations. For this plane condition, displacements can develop in any direction; the only plane restriction that must be satisfied is that all displacement derivatives with reference to the third direction are equal to zero.

Considering the governing equations for the three-dimensional case, given in chapter three, and enforcing the above assumptions, two different sets of equations are obtained. One of them represents the plane problem already discussed in the previous chapter. The second set is a potential problem and it is known as the "anti-plane problem". The mentioned resulting problems are independent from each other and their respective boundary equations can be formulated separately.

The boundary equations for the anti-plane case are obtained once more by applying the weighted residual technique, although their derivations are also possible directly from the corresponding three-dimensional expressions. The case of initial strain or stress type loads has been introduced in the formulation and the resulting derivatives of singular integral are again performed following the Mikhlin (43) concept.

4.2 Governing Equations and Fundamental Solution

For a body defined by the domain Ω and the boundary Γ

the governing equilibrium equations have already been presented in section(3.2). Two and three-dimensional problems were represented by the same set of equations and they were differentiated by the range of the indices involved in each particular case. To introduce the complete plane strain conditions, different ranges should be defined. The body (Ω, Γ) is still defined in the context of the two-dimensional space $(x_j, j = 1,2)$ but its associated displacements have three-dimensional representation $(u_i, i = 1,2,3)$, although they are functions of x_j with $j = 1,2$ only.

Considering that the displacement and stress conditions are identical in all planes perpendicular to the third axis (x_3) of the Cartesian system $(x_j, j = 1,2,3)$, the following assumptions are made to define the complete plane strain case:-

(a) Displacements have three-dimensional representation, but they are only functions of a two-dimensional Cartesian system of coordinates, i.e.,

$$u_i = f_i(x_1, x_2) \qquad (4.2.1)$$

in which i has a range of three.

(b) The partial derivative of a displacement (u_i) with respect to a coordinate (x_j) is only a function of the plane system,

$$\frac{\partial u_i}{\partial x_j} = f'_{ij}(x_1, x_2) \qquad (4.2.2)$$

when $j = 1,2$, and vanishes when $j = 3$, i.e.,

$$\frac{\partial u_i}{\partial x_3} = 0 \qquad (4.2.3)$$

The Navier equation for the three-dimensional case was given in section (3.2) and for easier comprehension is repeated here,

$$\frac{1}{(1-2\nu)} u_{j,ij} + u_{i,jj} + \frac{\bar{b}_i}{G} = 0 \qquad (4.2.4)$$

which is a set of three partial differential equations with $(i,j = 1,2,3)$.

Taking into account the assumptions made for the complete plane strain case, the above equation is transformed into two independent groups of partial differential equations. The first is the Navier equation for the usual plane strain, and is represented by the same equation (4.2.4), considering now ranges of two for i and j, i.e., $i,j = 1,2$. The second group is represented by the partial equilibrium equation related to the third direction and is given by,

$$u_{3,kk} + \frac{\bar{b}_3}{G} = 0 \qquad (4.2.5)$$

or when the Laplace operator is employed becomes,

$$\nabla^2(u_3) + \frac{\bar{b}_3}{G} = 0 \qquad (4.2.6)$$

with the equivalent body forces given as follows,

$$\bar{b}_3 = b_3 - \sigma^o_{3j,j} \qquad (4.2.7)$$

In a similar way the other equations in section (3.2) can also be written separately for the usual plane strain and anti-plane problems. All plane equations have already been given in section (3.2); here only the anti-plane expression for the traction vector representation is given, i.e,

$$\bar{p}_3 = G(u_{3,j})\eta_j \qquad (4.2.8)$$

in which

$$\bar{p}_3 = p_3 + \sigma^o_{3j} n_j \qquad (4.2.9)$$

The fundamental solution corresponding to equation (4.2.5) represents a response u^*_3 (displacement in the X_3 direction) at "q" (field point) due to a unit source applied at "s" (load point) in an infinite plane body (Ω^*, Γ^*). This solution can be obtained by replacing the body force term in equation (4.2.5) by the Dirac delta $\delta(s,q)$ as has been done in section (3.2) for the Navier equation. Thus, equation (4.2.5) becomes,

$$u^*_{3,kk} + \frac{\delta(s,q)}{6} = 0 \qquad (4.2.10)$$

which gives the sought fundamental solution (1,2).

An alternative procedure to obtain the solution is given by convenient integration of the three-dimensional Kelvin fundamental solution (eq. 3.2.28). Let us, therefore, consider the three-dimensional idealization of the infinite domain Ω^* (fig. 4.2.1) in which the displacement solution due to a unit point load applied in the third direction is represented by $u^*_{3k}(\bar{s},q)$. In order to obtain the anti-plane solution one has to compute the response of a line of unit loads by integrating them within the range $-\infty$ to ∞. Thus, the anti-plane fundamental solution is given by,

$$u^*_3(s,q) = \lim_{M \to \infty} \int_{-M}^{M} u^*_{33}(\bar{s},q) dx_3(\bar{s}) \qquad (4.2.11)$$

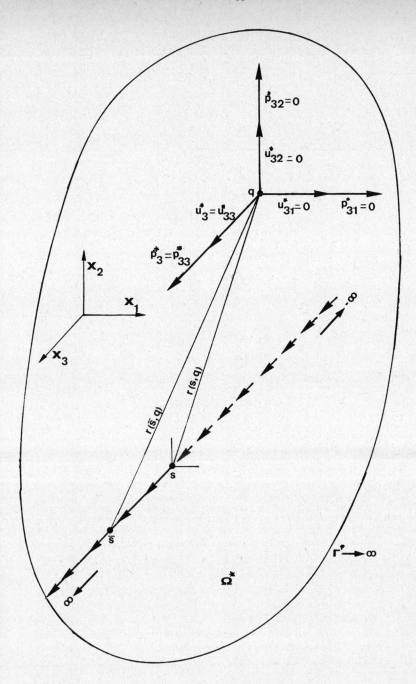

Figure 4.2.1 Fundamental Solution for the Anti-Plane Problem.

or

$$u_3^*(s,q) = \frac{1}{16\pi G(1-\nu)} \lim_{M\to\infty} \left[\int_{-M}^{M} (3-4\nu) \frac{1}{r(\bar{s},q)} dx_3(\bar{s}) + \int_{-M}^{M} \frac{r_{,3} r_{,3}}{r(\bar{s},q)} dx_3(\bar{s}) \right]$$
(4.2.12)

in which

$$r(\bar{s},q) = \left[(x_1(q)-x_1(\bar{s}))^2 + (x_2(q)-x_2(\bar{s}))^2 + (x_3(q)-x_3(\bar{s}))^2 \right]^{\frac{1}{2}}$$ (4.2.13)

After performing the limits in expression (4.2.12) an infinite value entirely consistent with the physical nature of two-dimensional problems is obtained. The infinite part can be interpreted as a rigid body motion which does not affect the relative displacements and can be neglected. Therefore, the fundamental solution is given by taking only the finite value as follows,

$$u_3^*(s,q) = -\frac{1}{2\pi G} \ell n r$$ (4.2.14)

The same result is obtained if the three-dimensional fundamental solution were modified by a particular constant value which also represents the introduction of a rigid body movement and removes the infinite term in the final expression.

In order to obtain the effects of the anti-plane unit load in the plane directions, one has to integrate the fundamental solution $u_{31}^*(\bar{s},q)$ and $u_{32}^*(\bar{s},q)$ over the same line parallel to the third direction used for $u_3^*(s,q)$ determination. As expected, these integrals vanish. Therefore, no displacement in the plane direction occurs, due to a load applied in the anti-plane problem. So, once more the independence between plane and anti-plane problems is demonstrated.

Using equation (4.2.14) the traction fundamental solution can be derived as follows,

$$p_3^*(s,q) = G \frac{\partial u_3^*}{\partial n}(s,q) \qquad (4.2.15)$$

or in an explicit form,

$$p_3^*(s,q) = -\frac{1}{2\pi}\frac{1}{r} r_{,k} n_k \qquad (4.2.16)$$

4.3 Integral Equations for Interior Points

The weighted residual technique can also be applied to derive integral equations referred to an interior point in the domain Ω of a bounded body for the potential problem represented by the anti-plane governing equation (4.2.5). Using the weighted residual statement for the potential theory (1) the following equation can be written,

$$\int_\Omega (\sigma_{3k,k}+b)u^* d\Omega = \int_{\Gamma_2}(p-\bar{p})u^* d\Gamma + \int_{\Gamma_1}(\bar{u}-u)p^* d\Gamma \qquad (4.3.1)$$

in which the values represented without any subscript are related to the third direction of the rectangular system of coordinates; u^* is a potential fundamental solution given in the preceding section; p^* is the normal derivative (see equation 4.2.15); u, p, $\bar{u}$ and $\bar{p}$ are the unknown and prescribed values of the potential and its derivatives; Γ_1 and Γ_2 represent parts of the boundary Γ where tractions and displacements are prescribed respectively.

Integrating equation (4.3.1) by parts and writing the expression in absence of the body forces gives,

$$-\int_\Omega \sigma_{3k}\varepsilon_{3k}^* d\Omega = -\int_{\Gamma_2}\bar{p}u^* d\Gamma - \int_{\Gamma_1}pu^* d\Gamma + \int_{\Gamma_1}(\bar{u}-u)p^* d\Gamma \qquad (4.3.2)$$

in which

$$\varepsilon_{3k}^* = \frac{1}{2}\frac{\partial u^*}{\partial x_k} \qquad (4.3.3)$$

The shear stresses σ_{3k} can be given by its elastic and initial stress parts. Thus,

$$\sigma_{3k} = \sigma_{3k}^e - \sigma_{3k}^o \tag{4.3.4}$$

Introducing (4.3.4) into (4.3.2) and integrating by parts once more gives,

$$-\int_\Omega \sigma_{3k,k}^* u\,d\Omega = -\int_\Gamma p^* u\,d\Gamma + \int_\Gamma u^* p\,d\Gamma + \int_\Omega \sigma_{3k}^o \varepsilon_{3k}^*\,d\Omega \tag{4.3.5}$$

Using equation (4.2.10) and the property of the Dirac Delta (Eq. 3.2.25), equation (4.3.5) becomes,

$$u(s) = -\int_\Gamma p^*(s,Q)u(Q)d\Gamma(Q) + \int_\Gamma u^*(s,Q)p(Q)d\Gamma(Q) + \int_\Omega \varepsilon_{3k}^*(s,q)\sigma_{3k}^o(q)d\Omega(q) \tag{4.3.6}$$

where

$$\varepsilon_{3k}^*(s,q) = -\frac{1}{4\pi G}\frac{1}{r} r,_k \tag{4.3.7}$$

Equation (4.3.6) gives the anti-plane displacement at an interior point "s". The shear stress representation of that problem can be obtained by its derivatives. Thus,

$$\sigma_{3j}(s) = G\frac{\partial}{\partial x_j(s)}\left[-\int_\Gamma p^*(s,Q)u(Q)d\Gamma(Q)\right] + G\frac{\partial}{\partial x_j(s)}\left[\int_\Gamma u^*(s,Q)p(Q)d\Gamma(Q)\right]$$

$$+ \frac{G}{\partial x_j(s)}\left[\int_\Omega \varepsilon_{3k}^*(s,q)\sigma_{3k}^o(q)d\Omega(q)\right] - \sigma_{3k}^o(s) \tag{4.3.9}$$

in which the first two derivatives can be easily performed,

$$\sigma_{3j}(s) = -\int_\Gamma S_{3j}(s,Q)u(Q)d\Gamma + \int_\Gamma D_{3j}(s,Q)p(Q)d\Gamma(Q) +$$

$$+ \frac{G\partial}{\partial x_j(s)}\left[\int_\Omega \varepsilon_{3k}^*(s,q)\sigma_{3k}^o(q)d\Omega(q)\right] - \sigma_{3k}^o(s) \tag{4.3.9}$$

The tensors $S_{3j}(s,Q)$ and $D_{3j}(s,Q)$ obtained by the derivatives of $p^*(s,Q)$ and $u^*(s,Q)$ respectively are given by,

$$S_{3j}(s,Q) = -\frac{G}{2\pi}\frac{1}{r^2}\left[2r,_j(r,_m\eta_m)-\eta_j\right] \quad (4.3.10)$$

$$D_{3j}(s,Q) = \frac{1}{2\pi}\frac{1}{r}r,_j \quad (4.3.11)$$

The remaining derivative has to be performed in a proper way by using the singular integral concept (43). Let us examine this term separately,

$$I = G\frac{\partial}{\partial x_j(s)}\int_\Omega \varepsilon^*_{3k}(s,q)\sigma^o_{3k}(q)d\Omega(q) \quad (4.3.12)$$

which can be written as follows,

$$I = G\int_\Omega \frac{\partial}{\partial x_j(s)}\left(\frac{e^*_k(s,q)}{r}\right)\sigma^o_{3k}(q)d\Omega(q) - G\sigma^o_{3k}(s)\int_0^{2\pi} e^*_k(s,\bar{q})r,_j d\theta(\bar{q}) \quad (4.3.13)$$

where

$$e^*_k(s,q) = \varepsilon^*_{3k}(s,q)r \quad (4.3.14)$$

Carrying out the derivative and performing the defined integral, the complete stress expression (4.3.8) becomes,

$$\sigma_{3j}(s) = -\int_\Gamma S_{3j}(s,Q)u(Q)d\Gamma(Q) + \int_\Gamma D_{3j}(s,Q)p(Q)d\Gamma(Q) +$$
$$+ \int_\Omega E_{3jk}(s,q)\sigma^o_{3k}(q)d\Omega(q) - \frac{1}{2}\sigma^o_{3\ell}\delta_{\ell j} \quad (4.3.15)$$

in which the tensor $E_{3jk}(s,q)$ is given by,

$$E_{3jk}(s,q) = -\frac{1}{4\pi r^2}\left[2r,_j r,_k -\delta_{kj}\right] \quad (4.3.16)$$

An analogous procedure can be developed in order to formulate the integral equation due to an initial strain applied in the domain. Thus, for an interior point "s" one can express displacements and stresses respectively as follows,

$$u(s) = -\int_\Gamma p^*(s,Q)u(Q)d\Gamma(Q) + \int_\Gamma u^*(s,Q)p(Q)d\Gamma(Q) \qquad (4.3.17)$$
$$+ \int_\Omega \sigma^*_{3k}(s,q)\varepsilon^o_{3k}(q)d\Omega(q)$$

$$\sigma_{3j}(s) = -\int_\Gamma S_{3j}(s,Q)u(Q)d\Gamma(Q) + \int_\Gamma D_{3j}(s,Q)p(Q)d\Gamma(Q) \qquad (4.3.18)$$
$$+ \int_\Omega F_{3jk}(s,q)\varepsilon^o_{3k}(q)d\Omega(q) - G\varepsilon^o_{3\ell}\delta_{j\ell}$$

in which

$$\sigma^*_{3k}(s,q) = -\frac{1}{2\pi}\frac{r,_k}{r} \qquad (4.3.19)$$

and

$$F_{3jk}(s,q) = 2GE_{3jk}(s,q) \qquad (4.3.20)$$

It is important to notice that the domain integrals in equations (4.3.15) and (4.3.18) have to be performed in the sense of principal Chauchy value, due to the order of singularities involved.

4.4 Boundary Integral Equation

The boundary integral equation to provide a necessary relation between surface displacements and tractions can be derived by taking an interior point "s" to the boundary. Analogously to the previous case, plane problem, that relation can be found by considering a boundary point "S" as an interior point "s" with the increase of the domain Ω to $\Omega+\Omega_\varepsilon$ (fig. 3.3.1). If the limit of the displacement equation is taken when $\Omega_\varepsilon \to 0$ (see ref.1), the final boundary equation for the anti-plane problem is obtained,

$$c(S)u(S) = -\int_\Gamma p^*(S,Q)u(Q)d\Gamma(Q) + \int_\Gamma u^*(S,Q)p(Q)d\Gamma(Q) \qquad (4.4.1)$$
$$+ \int_\Omega \varepsilon^*_{3k}(S,q)\sigma^o_{3k}(q)d\Omega(q)$$

in which c(s) is a scalar value also depending only on the geometric features of the boundary Γ as can be seen in Chapter 5.

Equation (4.4.1) was derived for bounded bodies but it can be easily demonstrated to be valid for infinite domains.

CHAPTER 5

BOUNDARY ELEMENT METHOD

5.1 Introduction

In this chapter a general procedure to obtain a numerical approach to solve the integral equations for plane (eqs. 3.3.5 and 3.3.7) and anti-plane (eqs. 4.3.15 and 4.4.1) cases previously formulated, is presented.

By considering the difficulties involved in solving the above mentioned equations with closed form solutions for almost all practical applications in continuum mechanics, the transformation of them into linear algebraic equations is shown. The transformation involves the discretization of the boundary of any body into elements over which displacements and tractions are interpolated according to chosen trial functions. Similarly, the domain integrals can be computed numerically over small areas called cells, over which interior values such as body forces, initial stresses, etc., are also calculated by interpolation functions.

The linear algebraic equations resulting from the transformation of boundary equations for plane (eq. 3.5.7) and anti-plane (4.4.1) problems constitute a linear system of equations in which the unknowns are represented by displacement and traction values. Although the system is formed by equations resulting from relations (3.5.7) and (4.4.1), modifications are shown to be needed for some cases. For instance, the analysis of a piecewise homogeneous body requires that the system of equations be modified by introducing equilibrium and displacement compatibility conditions.

In order to take into account the non-uniqueness of the traction vector at a boundary point, extra conditions must be introduced into the system of equations replacing some boundary relations. Another modification is shown to be necessary for the analysis of thin subregions, where all the integral equations for the subregion are replaced by other relations.

Finally, some practical applications to illustrate the theory developed so far are presented, and whenever possible the results are compared with analytical or other numerical solutions.

5.2 Discretization of the Integral Equations

The displacements at a boundary point "S" are related to all displacements and tractions over the boundary Γ through equations (3.3.5) and (4.4.1) for plane and anti-plane problems respectively. In these equations, the displacements at "S" are also affected by body forces and initial stress or strain term. In order to standardize the formulation to be presented, the above mentioned relations will be represented by one single form as follows,

$$\underline{c}(S)\underline{u}(S) + \int_{\Gamma} \underline{p}^*(S,Q)\underline{u}(Q)d\Gamma(Q) = \int_{\Gamma} \underline{u}^*(S,Q)\underline{p}(Q)d\Gamma(Q) +$$
$$+ \int_{\Omega} \underline{u}^*(S,q)\underline{b}(q)d\Omega(q) + \int_{\Omega} \underline{\varepsilon}^*(S,q)\underline{\sigma}^o(q)d\Omega(q) \qquad (5.2.1)$$

where $\underline{p}^*$ and $\underline{u}^*$ are matrices given by the traction and displacement fundamental solutions, $\underline{\varepsilon}^*$ is a matrix formed by strain components computed with the $\underline{u}^*$ terms, and $\underline{u}, \underline{p}, \underline{b}$ and $\underline{\sigma}^o$ are vectors for displacements, tractions, body forces and initial stresses. All matrices and vectors described have to be interpreted according to whether the plane or anti-plane case is under consideration.

For the discretization of equation (5.2.1) one has to use a procedure consisting of the following steps:-

(a)   The boundary Γ is approximated by a series of elements Γ_j (fig. 5.2.1). Displacements and tractions over each element Γ_j are assumed to be piecewisely approximated between nodal points, i.e., they are expressed in terms of interpolation functions ($\underline{\phi}$) and nodal values as follows,

$$\underline{u}^{(j)} = \underline{\phi}^T \underline{U}^{(N)}$$

$$\text{on } \Gamma_j \qquad (5.2.2)$$

$$\underline{p}^{(j)} = \underline{\phi}^T \underline{P}^{(N)}$$

in which $\underline{U}^{(N)}$ and $\underline{P}^{(N)}$ are vectors containing the nodal values associated with the element Γ_j, and $\underline{u}^{(j)}$ and $\underline{p}^{(j)}$ are the displacement and traction vectors for any point on the element Γ_j.

(b)   In order to perform the domain integrals indicated in equation (5.2.1), the domain Ω has been divided into cells (fig. 5.2.2). Then, body force and initial stress values, which are usually known only at some discrete points, are represented by function ($\underline{\psi}$) piecewisely defined over each cell Ω_m, i.e.,

$$\underline{b}^{(m)} = \underline{\psi}^T \underline{B}^{(N)}$$

$$\text{in } \Omega_m \qquad (5.2.3)$$

$$\underline{\sigma}^{o(m)} = \underline{\psi}^T \underline{\sigma}^{o(N)}$$

where $\underline{B}^{(N)}$ and $\underline{\sigma}^{o(N)}$ are vectors containing the known values of body forces and initial stresses at some particular points in Ω_m; $\underline{b}^{(m)}$ and $\underline{\sigma}^{o(m)}$ are the body force and initial stress vectors for any point in the cell Ω_m.

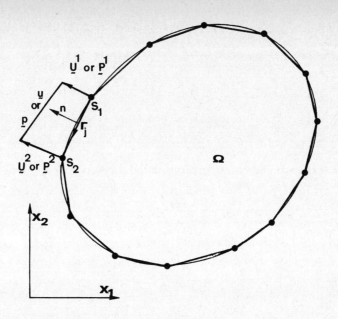

Figure 5.2.1 Boundary Discretizations.

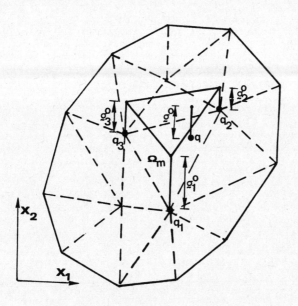

Figure 5.2.2 Internal Cells.

(c) Similarly the cartesian coordinate representation $\underline{x}$ for any point, whether on the boundary or in the domain can also be expressed in terms of interpolation functions and nodal values i.e.,

$$\underline{x}^{(j)} = \underline{\Phi}_c^T \underline{X}^{(N)} \quad \text{on} \quad \Gamma_j$$

and (5.2.4)

$$\underline{x}^{(m)} = \underline{\Phi}_c^T \underline{X}^{(N)} \quad \text{in} \quad \Omega_m$$

Using an approximation as given in equations (5.2.2), (5.2.3) and (5.2.4) the integral over both elements and cells can be convinient evaluated, and equation 5.2.1 for a boundary node "S" can be written in a discretized form as (1,2)

$$\underline{C}\underline{U} + \hat{\underline{H}}\underline{U} = \underline{G}\underline{P} + \underline{D}\underline{B} + \underline{E}\underline{\sigma}^o \qquad (5.2.5)$$

in which $\underline{U}$ and $\underline{P}$ contain all nodal displacements and tractions, $\underline{B}$ and $\underline{\sigma}^o$ represent the necessary known values of body forces and initial stresses in the domain, and $\hat{\underline{H}}$, $\underline{G}$, $\underline{D}$ and $\underline{E}$ are matrices computed by the integrals over boundary elements and cells indicated in equation (5.2.1).

Incorporating the matrix $\underline{C}$ into $\hat{\underline{H}}$, the system of equation can also be written as,

$$\underline{H}\underline{U} = \underline{G}\underline{P} + \underline{D}\underline{B} + \underline{E}\underline{\sigma}^o \qquad (5.2.6)$$

The above equation relates nodal displacements and tractions at defined nodes in the case of a plane problem. For a well posed problem with N nodes, the system (5.2.6) represents 2N linear algebraic equations in which N_1 displacements and N_2 tractions such as $2N = N_1 + N_2$ are known. Equation (5.2.6) also governs the relations between displacements and tractions for the anti-plane case. The system is now composed of only N equations corresponding to N nodal points defined on the boundary. In this case, the total number of prescribed displacements (N_1) and tractions (N_2) has to be equal to N. For instance, the product $\underline{DB}$ in equation (5.2.6) must be neglected for the anti-plane case, according to the formulation presented in chapter 4.

Following the same procedure the displacement and stress representations at an interior point (equations 3.3.1 and 3.3.7 for plane problems, and equations 4.3.6 and 4.3.15 for anti-plane problems) can also be expressed in matrix forms as follows,

$$\bar{\underline{U}} = - \underline{HU} + \underline{GP} + \underline{DB} + \underline{E}\sigma^o \qquad (5.2.7)$$

and

$$\underline{\sigma} = - \underline{H'U} + \underline{G'P} + \underline{D'B} + (\hat{\underline{E}} + \bar{\underline{E}})\sigma^o \qquad (5.2.8)$$

or

$$\underline{\sigma} = - \underline{H'U} + \underline{G'P} + \underline{D'B} + \underline{E'}\sigma^o \qquad (5.2.9)$$

with

$$\underline{E'} = \underline{\hat{E}} + \underline{\bar{E}} \qquad (5.2.10)$$

where $\bar{\underline{U}}$ and $\underline{\sigma}$ are vectors containing the displacements and stresses of an interior point; $\underline{H}$, $\underline{S}$, $\underline{D}$, $\underline{E}$, $\underline{H}'$, $\underline{G}'$, $\underline{D}'$, $\underline{E}'$ are matrices obtained after performing the integral over all boundary elements and internal cells.

Equations (5.2.8) and (5.2.10) were formulated by internal points only, however a similar representation is obtained for boundary points (see Appendix A), by taking into account tractions and derivatives of displacements over the boundary elements connected with the node under consideration.

All terms of the matrices presented in equations (5.2.6), (5.2.7) and (5.2.8) are obtained by numerical or analytical integrals over the boundary elements or internal cells as already presented elsewhere (38). However some of these terms are recomended here to be evaluated by mea s of analytical expression due to the type of singularities envolved.

The leading diagonal terms of $\underline{H}$ is usually computed using the rigid body translation concept (1,2), but in many instances the actual evaluation of the matrix $\underline{C}$ (equation 3.3.6) and the proper determination of the first integral of equation (5.2.1) computed in the Chauchy principal value sense is recomended.

The coefficients of $\underline{C}$ can be determined by performing the integral over the boundary Γ_ε of a small ball (fig. 5.2.3) and then taking the limit when $\varepsilon \to 0$. The $\underline{C}$ coefficients at a boundary node "S" are then given by,

$$\underline{C}(S) = \underline{\beta} + \lim_{\varepsilon \to 0} \int_{\Gamma_\varepsilon} \underline{p}^*(S,Q) d\Gamma(Q) \qquad (5.2.11)$$

where β represents a unit (2 x 2) matrix or a unit scalar value depending on whether the plane or anti-plane problem has been considered.

Substitutint the element of boundary $d\Gamma(Q)$ the corresponding element of angle $d\theta$, i.e.,

$$d\Gamma(Q) = \varepsilon d\theta \qquad (5.3.12)$$

the integrals over the angle θ from θ_o to θ_1 become independent of ε, resulting in final integrated values which are functions of the angles of the elements connected at "S". The final components of $\underset{\sim}{C}$ can be expressed as function of two angles, α and γ, defined in figure (5.3.3). For plane case, they are given by,

$$\underset{\sim}{C} = \begin{bmatrix} \dfrac{\alpha}{2\pi} + \dfrac{\cos(2\gamma)\sin\alpha}{4\pi(1-\nu)} & \dfrac{\sin(2\gamma)\sin\alpha}{4\pi(1-\nu)} \\ \\ \dfrac{\sin(2\gamma)\sin\alpha}{4\pi(1-\nu)} & \dfrac{\alpha}{2\pi} - \dfrac{\cos(2\gamma)\sin\alpha}{4\pi(1-\nu)} \end{bmatrix} \qquad (5.2.13)$$

and for the anti-plane case,

$$\underset{\sim}{C} = \dfrac{\alpha}{2\pi} \qquad (5.2.14)$$

In order to perform the integral over all elements to compute the matrices $\underset{\sim}{H}$ and $\underset{\sim}{G}$, an interpolation function to approximate displacements $\underset{\sim}{u}$ and tractions $\underset{\sim}{p}$ must be chosen. Linear variation over the length of each element is the only case applied throughout this work. Although the linear element has been found to give acceptable accuracy for the applications presented here, higher order approximations can also be formulated using an analogous procedure.

The procedure to compute the other terms for all matrices involued in equation (5.2.6) requires the definition of the interpolation functions to approximate tractions displacements, body forces, initial stresses and coordinates as shown in equations (5.2.2), (5.2.3) and (5.2.4). Using linear approximation over a boundary element Γ_j as shown in figure (5.2.4) tractions and displacements can be represented as follows,

$$\underset{\sim}{u}^{(j)} = \begin{bmatrix} u_1 \\ u_2 \end{bmatrix} = \begin{bmatrix} \phi^1 & \phi^2 & 0 & 0 \\ 0 & 0 & \phi^1 & \phi^2 \end{bmatrix} \begin{bmatrix} u_1^{(1)} \\ u_1^{(2)} \\ u_2^{(1)} \\ u_2^{(2)} \end{bmatrix}$$

$$\underset{\sim}{p}^{(j)} = \begin{bmatrix} p_1 \\ p_2 \end{bmatrix} = \begin{bmatrix} \phi^1 & \phi^2 & 0 & 0 \\ 0 & 0 & \phi^1 & \phi^2 \end{bmatrix} \begin{bmatrix} p_1^{(1)} \\ p_1^{(2)} \\ p_2^{(1)} \\ p_2^{(2)} \end{bmatrix}$$

(5.2.15)

for the plane case, and

$$\underset{\sim}{u}^{(j)} = \begin{bmatrix} u_3 \end{bmatrix} = \begin{bmatrix} \phi^1 & \phi^2 \end{bmatrix} \begin{bmatrix} u_3^{(1)} \\ u_3^{(2)} \end{bmatrix}$$

$$\underset{\sim}{p}^{(j)} = \begin{bmatrix} p_3 \end{bmatrix} = \begin{bmatrix} \phi^1 & \phi^2 \end{bmatrix} \begin{bmatrix} p_3^{(1)} \\ p_3^{(2)} \end{bmatrix}$$

(5.2.16)

for the anti-plane case

In expressions (5.2.15) and (5.2.16) the superscript indices 1 and 2 are related to the element nodes, the subscripts indicate displacement

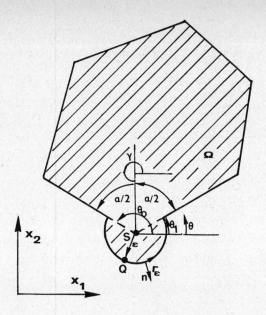

Figure 5.2.3 Definition of $\underset{\sim}{C}$ Terms.

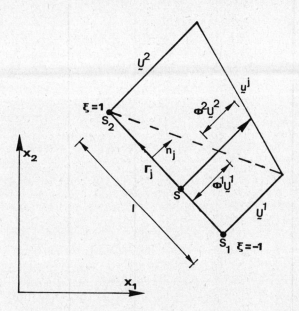

Figure 5.2.4 Linear Element.

and traction directions, and the interpolation functions, Φ^1 and Φ^2, are given by,

$$\Phi^1 = -\frac{1}{2}(\xi-1)$$

$$\Phi^2 = \frac{1}{2}(\xi+1)$$

(5.2.17)

Dividing the boundary of the body under consideration into elements and then substituting equation (5.2.15) or (5.2.16) into the integral terms allow us to obtain submatrices $\underset{\sim}{h}$ and $\underset{\sim}{g}$ dependent only on the fundamental solution interpolation function and boundary geometry as follows,

$$h_{ik}^{mn} = \frac{\ell}{2}\int_{-1}^{1} p_{ik}^{*} \, \Phi^m \, d\xi$$

$$g_{ik}^{mn} = \frac{\ell}{2}\int_{-1}^{1} u_{ik}^{*} \, \Phi^m \, d\xi$$

(5.2.18)

where n indicates the singular node, and ξ is the dimensionless coordinate over the boundary.

Equation (5.2.18) is given in tensorial notation and is valid only for plane case. For anti-plane problems the same equation can be written, but the subscripts i and k must be neglected.

As has been said previously, the integration of these expressions can be performed by numerical means when the singular point is not on the boundary element without any loss of accuracy in the results. Since the singular point is on the element under consideration, analytical integration is preferred as more sophisticated schemes are necessary for numerical integration. For this case the values of the submatrices $\underline{g}$ obtained analytically from the indicated integrals are expressed by,

$$g_{ik}^{mn} = \frac{\ell}{16\pi G(1-\nu)} \left[-(3-4\nu)(\log \ell - \frac{1}{2} - \delta_{mn}) + \eta_i \eta_k \right] \quad (5.2.19)$$

for the plane case.

The g^{mn} terms for the anti-plane case can be analogously represented by,

$$g^{mn} = -\frac{1}{4\pi G}(\log \ell - \frac{1}{2} - \delta_{mn}) \quad (5.2.20)$$

It should be noticed that the analytical integration over a linear element can be avoided by employing numerical schemes, such as one-dimensional Gaussian quadrature (24), but it has been observed in practical applications that a considerable number of integration points must be used for reasonable accuracy in the integrated values.

Analytical integration over the element can also be used to obtain the values of h_{ik}^{mn} for $m \neq n$, i.e., when the singular point is the node where the interpolation function is zero. Then for the plane case one has,

$$h_{ik}^{mn} = \frac{\ell}{8\pi(1-\nu)}(i-k)(1-\delta_{mn})(m-n) \quad (m \neq n) \quad (5.2.21)$$

and for the anti-plane case

$$h^{mn} = 0 \quad (5.2.22)$$

For cases where m and n assume the same value, the integral must be carried out in the Cauchy principal value sense. The integral over the two elements connected to the node must be performed together in order to eleminate the singularity which arises when the limit of each integral is enforced. No integrated value h_{ik}^{mn} (m=n) can be expressed individually, however the total value of each component in the submatrix $\underset{\sim}{H}_{ik}$ corresponding to the node under consideration can be calculated for plane cases as follows,

$$H_{ik} = \frac{(1-2\nu)}{4\pi(1-\nu)} \log(\bar{\ell}/\ell)(k-i) \qquad (5.2.23)$$

in which $\bar{\ell}$ is the length of the second element connected to the node following the orientation of boundary .

For the anti-plane case, zero values of the principal integral are obtained with the same procedure.

The determination of integrals over the boundary elements considering internal load points for the determination of displacements and stresses does not involve any singularity and numerical schemes are used without significant loss of accuracy in the integrated values.

Similarly the domain integrals for displacement or stress determination can be expressed as integrals over cells and after introducing interpolation functions submatrices as in equation (5.2.18) can be defined. In this work the corresponding submatrices for each cell is obtained following a scheme proposed in reference (38) in which the terms are obtained using a semi-analytical procedure. By using a cylindrical system of coordinates the analytical integral over one dimension is very simple and one has to compute numerically, only the integral over the angular coordinate.

5.3 Subregions

The matrix equations (5.2.6), (5.2.7) and (5.2.9) were formulated for a homogeneous body (with finite or infinite boundary). These equations can now be extended for the analysis of piecewise homogeneous bodies, i.e., structural system formed by several homogeneous subregions with possible different elastic properties. In order to deal with this problem the subregion technique (60,98), which consists of applying equation (5.2.6) for each subregion separately and then adding them together using equilibrium and displacement compatibility conditions, is adopted. So, for a domain formed by several subregions Ω_i (fig. 5.3.1), the matrix equation (5.2.6) can be written for all subregions in the following form,

$$\underset{\sim}{H}^{(i)} \underset{\sim}{U}^{(i)} = \underset{\sim}{G}^{(i)} \underset{\sim}{P}^{(i)} + \underset{\sim}{D}^{(i)} \underset{\sim}{B}^{(i)} + \underset{\sim}{E}^{(i)} \sigma^{o(i)} \qquad (5.3.1)$$

where the summation notation is not implied.

Considering the interface Γ_{ij} (common boundary for subregions "i" and "j", as can be seen in fig. 5.3.1), displacement compatibility and equilibrium conditions are given respectively by,

$$\underset{\sim}{U}^{(ij)} = \underset{\sim}{U}^{(ji)} \qquad (5.3.2)$$

and

$$\underset{\sim}{P}^{(ij)} = - \underset{\sim}{P}^{(ji)} \qquad (5.3.3)$$

in which the first superscript represents the subregion to which the displacement and traction are related, and the second index is the adjacent subregion number.

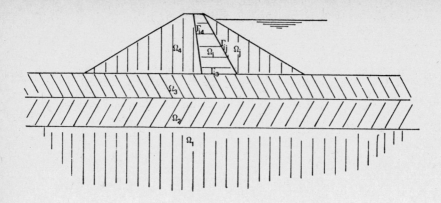

Figure 5.3.1 Subregions.

Vectors $\underline{U}^{(ij)}$ and $\underline{P}^{(ij)}$ represent boundary values over the interface. As some external boundary values are also related to subregion "i", the complete vectors $\underline{U}^{(i)}$ and $\underline{P}^{(i)}$ can be written divided in two parts, as follows,

$$\underline{U}^{(i)} = \begin{bmatrix} \underline{U}^{(ie)} \\ \underline{U}^{(ij)} \end{bmatrix} \qquad (5.3.4)$$

and

$$\underline{P}^{(i)} = \begin{bmatrix} \underline{P}^{(ie)} \\ \underline{P}^{(ij)} \end{bmatrix} \qquad (5.3.5)$$

where $\underline{U}^{(ie)}$ and $\underline{P}^{(ie)}$ stand for external boundary values.

Equation (5.3.1) can now be written as,

$$\begin{bmatrix} \underset{\sim}{H}^{(ie)} & \underset{\sim}{H}^{(ij)} \end{bmatrix} \begin{bmatrix} \underset{\sim}{U}^{(ie)} \\ \underset{\sim}{U}^{(ij)} \end{bmatrix} = \begin{bmatrix} \underset{\sim}{G}^{(ie)} & \underset{\sim}{G}^{(ij)} \end{bmatrix} \begin{bmatrix} \underset{\sim}{P}^{(ie)} \\ \underset{\sim}{P}^{(ij)} \end{bmatrix} + \underset{\sim}{D}^{(i)} \underset{\sim}{B}^{(i)} + \underset{\sim}{E}^{(i)} \underset{\sim}{\sigma}^{o(i)}$$

(5.3.6)

in which $\underset{\sim}{H}^{(ie)}$ and $\underset{\sim}{G}^{(ie)}$ represent the coefficients related to external boundary.

Similar expressions can also be written for subregion "j",

$$\begin{bmatrix} \underset{\sim}{H}^{(je)} & \underset{\sim}{H}^{(ji)} \end{bmatrix} \begin{bmatrix} \underset{\sim}{U}^{(je)} \\ \underset{\sim}{U}^{(ji)} \end{bmatrix} = \begin{bmatrix} \underset{\sim}{G}^{(je)} & \underset{\sim}{G}^{(ji)} \end{bmatrix} \begin{bmatrix} \underset{\sim}{P}^{(je)} \\ \underset{\sim}{P}^{(ji)} \end{bmatrix} + \underset{\sim}{D}^{(j)} \underset{\sim}{B}^{(j)} + \underset{\sim}{E}^{(j)} \underset{\sim}{\sigma}^{o(j)}$$

(5.3.7)

Equations (5.3.6) and (5.3.7) can now be added together using expressions (5.3.2) and (5.3.3) to give,

$$\begin{bmatrix} \underset{\sim}{H}^{(ie)} & \underset{\sim}{H}^{(ij)} & -\underset{\sim}{G}^{(ij)} & 0 \\ 0 & \underset{\sim}{H}^{(ji)} & \underset{\sim}{G}^{(ji)} & \underset{\sim}{H}^{(je)} \end{bmatrix} \begin{bmatrix} \underset{\sim}{U}^{(ie)} \\ \underset{\sim}{U}^{(ij)} \\ \underset{\sim}{P}^{(ij)} \\ \underset{\sim}{U}^{(je)} \end{bmatrix} = \begin{bmatrix} \underset{\sim}{G}^{(ie)} & \underset{\sim}{G}^{(ij)} & 0 & 0 \\ 0 & 0 & \underset{\sim}{G}^{(je)} & \underset{\sim}{G}^{(ji)} \end{bmatrix} \times$$

$$\begin{bmatrix} \underset{\sim}{P}^{(ie)} \\ \underset{\sim}{\bar{P}}^{(ij)} \\ \underset{\sim}{P}^{(ie)} \\ \underset{\sim}{\bar{P}}^{(ji)} \end{bmatrix} + \begin{bmatrix} \underset{\sim}{D}^{(i)} & 0 \\ 0 & \underset{\sim}{D}^{(j)} \end{bmatrix} \begin{bmatrix} \underset{\sim}{B}^{(i)} \\ \underset{\sim}{B}^{(j)} \end{bmatrix} + \begin{bmatrix} \underset{\sim}{E}^{(i)} & 0 \\ 0 & \underset{\sim}{E}^{(j)} \end{bmatrix} \begin{bmatrix} \underset{\sim}{\sigma}^{o(i)} \\ \underset{\sim}{\sigma}^{o(j)} \end{bmatrix}$$

(5.3.8)

in which $\underset{\sim}{P}^{(ij)}$ represents prescribed load on the interface.

Equation (5.3.8) is the boundary equation taking into account the interface between subregion "i" and "j". It can also be represented by the usual matrix boundary equation.

$$HU = GP + DB + E\sigma^o \qquad (5.3.9)$$

Similar procedure is used to provide stress and displacement equations for the piecewise body given in figure (5.3.1). Also in this case, the original matrix representations remain unchanged, i.e.,

$$\sigma = -H'U + G'P + D'B + E'\sigma^o \qquad (5.3.10)$$

$$\bar{U} = HU + GP + DB + E\sigma^o \qquad (5.3.11)$$

In these expressions the stress and displacement values are only dependent on coefficients relating to the subregion which contains the point under consideration.

5.4 Traction Discontinuities

In this section the non-uniqueness of the traction vector at a boundary node of a region or subregion will be reexamined. The formulation for this problem has already been presented in reference (67) in which the double point concept has been used. Here, the technique is extended in order to take into account the initial stress term which must be introduced to consider temperature or similar loads.

For a better understanding of the problem, let us consider the adjacent boundary elements (fig. 5.4.1) where nodes "Q" and "S" are defined with the same coordinates, and each of them is connected with only one element.

A simple condition enforcing the same displacement at "Q" and "S" can provide the necessary constraints (or constraint for the anti-plane case) in order to solve the problem, i.e.,

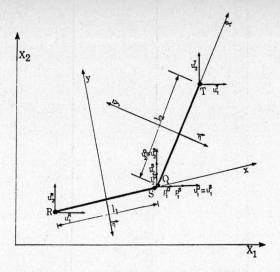

Figure 5.4.1 Double Nodes.

$$u_k^{(Q)} = u_k^{(S)} \qquad (5.4.1)$$

Although equation (5.4.1) is suitable for several traction discontinuity cases, other conditions (or condition for the anti-plane problem) must be used when either the double nodes are on an interface or the same displacement u_k is prescribed for "Q" and "S".

Assuming that the stress tensor is uniquely defined at the double nodes (Q and S) and writing the traction vector for the adjacent elements as a function of the stresses, the following condition is derived for the plane case,

$$p_1^{(S)} n_1^{(2)} - p_1^{(Q)} n_1^{(1)} + p_2^{(S)} n_2^{(2)} - p_2^{(Q)} n_2^{(1)} = 0 \qquad (5.4.2)$$

in which the superscripts indicate either nodes (S or Q) or elements (1 or 2), and the subscript indices refer to both traction directions and direction cosines.

Another condition for the plane case is derived enforcing the invariance of the strain trace tensor. Using linear interpolation function to approximate displacements over the two adjacent elements, the following condition can be written,

$$\left[(u_1^{(R)}-u_1^{(S)})n_2^{(1)}+(u_2^{(S)}-u_2^{(R)})n_1^{(1)}\right]/\ell_1 + \left[(u_1^{(T)}-u_1^{(Q)})n_2^{(2)} - (u_2^{(T)}-u_2^{(Q)})n_1^{(2)}\right]/\ell_2 - \frac{1}{2G}\left[p_1^{(Q)}n_1^{(2)}+p_2^{(Q)}n_2^{(2)}-p_1^{(S)}n_1^{(1)}-p_2^{(S)}n_2^{(1)}\right] =$$

$$= \frac{1}{2G}\left[\sigma_{11}^{o(Q)}n_2^{(2)}n_2^{(2)}-2\sigma_{12}^{o(Q)}n_2^{(2)}n_1^{(2)}+\sigma_{22}^{o(Q)}n_1^{(2)}n_1^{(2)}-\sigma_{11}^{o(S)}n_2^{(1)}n_2^{(1)}\right.$$

$$\left. + 2\sigma_{12}^{o(S)}n_1^{(1)}n_2^{(1)}-\sigma_{22}^{o(S)}n_1^{(1)}n_1^{(1)}\right] \qquad (5.4.3)$$

For anti-plane problems only one extra condition needs be introduced. This condition can be obtained by enforcing the continuity of the anti-plane shear strain at the double nodes. Expressing the shear strain in terms of displacements (at R,Q,S and T) and traction (at S), the following relation can be derived,

$$\left[u_3^{(Q)}-u_3^{(R)}\right]/\ell_1 - \left[u_3^{(T)}-u_3^{(S)}\right]\left[n_2^{(2)}n_2^{(1)}+n_1^{(2)}n_1^{(1)}\right]/\ell_2$$

$$= \frac{1}{G}(p_3^{(s)}+\sigma_{23}^{o(S)})(n_2^{(2)}n_1^{(1)}-n_1^{(2)}n_2^{(1)}) \qquad (5.4.4)$$

Equations (5.4.2), (5.4.3) and (5.4.4) have to be introduced into the system (5.3.9) conveniently replacing boundary equations computed for "S" or "Q". This procedure removes the singularity in the matrix $\underline{H}$ which is introduced when only boundary equations are used.

5.5 Thin Subregions

The subregion technique presented in Section (5.3) has proved capable of solving many structural systems formed of several homogeneous parts with different material properties. The technique can also be applied to solve the inclusion of a thin layer or seam, usually formed by soft soil, into rock masses. However in this situation, the interfaces defined by rock and seam materials have to be discretized into elements small enough in order to ensure that nodes on opposite interfaces are at a sufficient distance from each other in comparison with the size of the elements. As a result the final system of equations obtained is inconveniently large.

An alternative scheme to improve the boundary element formulation for such a problem consists of finding different expressions to replace all boundary integral equations related to the thin subregion. The seam interfaces have to be discretized into elements in such a way that opposite nodes are conveniently placed on the same orthogonal (when possible) as illustrated in figure (5.5.1). The rectangular domains formed by two opposite interface elements can be interpreted as seam elements over which one-dimensional stress-strain relations for shear and compression are assumed. Taking into account that the thickness of the seam, h_s, defined by nodes "S" and "S_1" (fig. 5.5.1) is very small in comparison with lengths, ℓ_1 and ℓ_2, of adjacent elements, assumption that the strain tensor does not vary over h_s is valid and its components at a boundary node "S" are given by,

$$\bar{\varepsilon}_{22}^{(S)} = \frac{\bar{u}_2^{(S_1)} - \bar{u}_2^{(S)}}{h_s} \qquad (5.5.1)$$

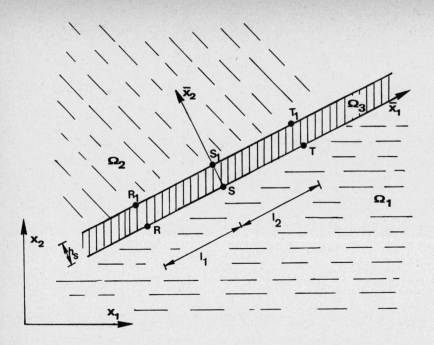

Figure 5.5.1 thin Subregion.

$$\bar{\varepsilon}_{12}^{(S)} = \frac{\bar{u}_1^{(S_1)} - \bar{u}_1^{(S)}}{h_s} \qquad (5.5.2)$$

for plane problems, and

$$\bar{\varepsilon}_{23}^{(S)} = \frac{u_3^{(S_1)} - u_3^{(S)}}{h_s} \qquad (5.5.3)$$

for anti-plane problems.

Using the above equation, the stresses at the boundary node "S" for the plane case are given by,

$$\bar{\sigma}_{22}^{(S)} = \frac{\bar{u}_2^{(S_1)} - \bar{u}_2^{(S)}}{h_s} E_s \qquad (5.5.4)$$

$$\bar{\sigma}_{12}^{(S)} = \frac{\bar{u}_1^{(S_1)} - \bar{u}_1^{(S)}}{h_s} G_s \qquad (5.5.5)$$

and for anti-plane case,

$$\bar{\sigma}_{23}^{(S)} = \frac{u_3^{(S_1)} - u_3^{(S)}}{h_s} G_s \qquad (5.5.6)$$

in which neither initial stress or initial strain terms have been considered.

For nodes "S" and "S_1", one needs four conditions (two for anti-plane case) to replace the usual boundary equations. Two of them are directly obtained by applying equilibrium condition, i.e.,

$$p_1^{(S_1)} + p_1^{(S)} = 0 \qquad (5.5.7)$$

$$p_2^{(S_1)} + p_2^{(S)} = 0 \qquad (5.5.8)$$

already referred to the global system of coordinates.

For the anti-plane case only one equation is written using equilibrium condition,

$$p_3^{(S_1)} + p_3^{(S)} = 0 \qquad (5.5.9)$$

The remaining equations are obtained by the relations between traction and stresses at node "S". Then,

$$p_2^{(S)} = -\bar{\sigma}_{22}^{(S)} \qquad (5.5.10)$$

$$p_1^{(S)} = -\sigma_{12}^{(S)} \qquad (5.5.11)$$

for plane problems, and

$$\bar{p}_3^{(S)} = -\bar{\sigma}_{23}^{(S)} \qquad (5.5.12)$$

for anti-plane problems.

A final form of the alternative conditions, already written in terms of the global system of coordinates, is derived by replacing the stress terms in equations (5.5.10) (5.5.11) and (5.5.12) by their values given in (5.5.4), (5.5.5) and (5.5.6) as follows.

$$\frac{G_s}{h_s} \begin{bmatrix} -n_2 & n_1 & n_2 & -n_1 \\ -n_1 & -n_2 & n_1 & n_2 \end{bmatrix} \begin{bmatrix} u_1^{(S)} \\ u_2^{(S)} \\ u_1^{(S_1)} \\ u_2^{(S_1)} \end{bmatrix} = \begin{bmatrix} -\dfrac{n_2}{2(1-\nu)} & \dfrac{n_1}{2(1-\nu)} \\ -n_1 & -n_2 \end{bmatrix} \begin{bmatrix} p_1^{(S)} \\ p_2^{(S)} \end{bmatrix} \qquad (5.5.13)$$

$$\frac{G_s}{h_s} \begin{bmatrix} 1 & -1 \end{bmatrix} \begin{bmatrix} u_3^{(S)} \\ u_3^{(S_1)} \end{bmatrix} = p_3^{(S)} \qquad (5.5.14)$$

for plane and anti-plane cases respectively.

Equations (5.5.7, 5.5.8, 5.5.9, 5.5.13, and 5.5.14) formulated above are the necessary conditions to replace the boundary integral equations for nodes "S" and "S_1".

5.6 Solution Technique

Although several modifications have been introduced into the boundary system of equations in sections (5.3 to 5.5), its initial form still remains for both plane and anti-plane problems,

$$\underline{H}\underline{U} = \underline{G}\underline{P} + \underline{D}\underline{B} + \underline{E}\underline{\sigma}^o \qquad (5.6.1)$$

Similarly the stress equations are represented by,

$$\underline{\sigma} = \underline{H}'\underline{U} + \underline{G}'\underline{P} + \underline{D}'\underline{B} + \underline{E}'\underline{\sigma}^o \qquad (5.6.2)$$

In order to improve the numerical procedure to solve any structural system, some of the indicated matrix operations to take into account boundary load (displacement and traction) and body force effects can be performed analytically, leading to final forms of expressions (5.6.1) and (5.6.2) as functions of the initial stress terms only. Then for a well-posed problem, with a sufficient number of boundary tractions or displacements prescribed, a final matrix form of equation (5.6.1) is obtained by interchanging columns in the matrices $\underline{H}$ and $\underline{G}$, in order to accumulate all unknowns in the vector $\underline{X}$ on the left hand side as follows,

$$\underline{A}\underline{X} = \underline{F} + \underline{E}\underline{\sigma}^o \qquad (5.6.3)$$

The vector $\underline{F}$ gives the effects of applied tractions, displacements and body forces.

The system of equation (5.6.3) consists of a non-symmetric matrix which may be banded or fully populated depending on whether subregions are considered or not. The matrix is usually well conditioned with dominant diagonal terms.

The solution of (5.6.3) is obtained by multiplying it by $\underline{A}^{-1}$, as follows,

$$\underline{X} = \underline{M} + \underline{R}\underline{\sigma}^o \qquad (5.6.4)$$

where

$$\underline{R} = \underline{\bar{A}}^{-1}\underline{E} \qquad (5.6.5)$$

and

$$\underline{M} = \underline{\bar{A}}^{-1}\underline{F} \qquad (5.6.6)$$

Using the same procedure in equation (5.6.2) one obtains,

$$\underline{\sigma} = \underline{F}' - \underline{A}'\underline{X} + \underline{E}\underline{\sigma}^o \qquad (5.6.7)$$

The above equation gives the total stress as a function of an initial stress field $\underline{\sigma}^o$. As we shall see in following chapters for many nonlinear applications, it is better to write,

$$\underline{\sigma}^e = \underline{F}' - \underline{A}'\underline{X} + \underline{E}*\underline{\sigma}^o \qquad (5.6.8)$$

where

$$\underline{E}* = \underline{E}' - \underline{I} \qquad (5.6.9)$$

and $\underline{\sigma}^e$ is defined as the elastic part of $\underline{\sigma}$, i.e.,

$$\underline{\sigma}^e = \underline{\sigma} - \underline{\sigma}^o \qquad (5.6.10)$$

Substituting (5.6.4) into (5.6.8) gives,

$$\underline{\sigma}^e = \underline{N} + \underline{S}\underline{\sigma}^o \qquad (5.6.11)$$

in which

$$\underline{S} = \underline{E}* - \underline{A}'\underline{R} \qquad (5.6.12)$$

and

$$\underline{N} = \underline{F}' - \underline{A}'\underline{M} \qquad (5.6.13)$$

It should be noticed that vectors $\underset{\sim}{M}$ and $\underset{\sim}{N}$ represent the solution of a problem without initial stresses, whose effects, when present, are modelled by matrices $\underset{\sim}{R}$ and $\underset{\sim}{S}$.

5.7 Practical Application of Boundary Element on Linear Problems

In this section geomechanical applications of the boundary element formulation for elastic problems are presented. Four examples are solved to illustrate some particular aspects of the technique formulated so far. In the first example the body force effects are computed using the equivalent boundary integrals presented in section (3.4). The subregion technique is illustrated in the second example in which a thin circular lining is adopted as the support of a pressure tunnel. The third application is concerned with a practical case of the complete plane strain problem. Finally, the thin layer formulation shown in section (5.5) is adopted to model a tunnel excavation.

(a)   Analysis of an embankment.

In this example a an embankment is analysed using linear elastic material. The problem was taken from reference (17) where the corresponding finite element solution is presented. The load applied to the structural system is the self-weight of the soil material, and its effects were computed using boundary integral instead of the domain integral, to avoid the definition of integral cells. The boundary was discretized into 20 linear elements and 23 nodes as is shown in fig. (5.7.1). In this case only half of the body has to be discretized in order to compute all matrices involved in the solution of the problem in their reduced form.

The displacement profile of the upper boundary of the dam (fig. 5.7.2) compares well with the finite element results, and stresses over the central line given in fig. (5.7.3) are almost linear with little perturbation due to the distant slopes.

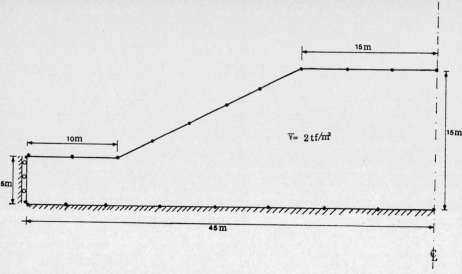

Figure 5.7.1 Embankment Discretization.

(b) Lined circular tunnel

This application consists of analysing a circular lined tunnel built in an infinite rock domain. The reinforced concrete lining adopted is 30 cm thick and its internal diameter is 400 cm as shown in figure (5.7.4). The applied load is an internal water pressure taken equal to 100 tf/m^2. Competent rock material is assumed to extend from the external lining surface to the infinity. This material is also divided into subregions defining a 115 cm thick ring around the circular lining.

Poisson's ratios for concrete and rock material were assumed 0.15 and 0.20 respectively, while the ratio between the elastic modulus of concrete and rock is equal to 2.

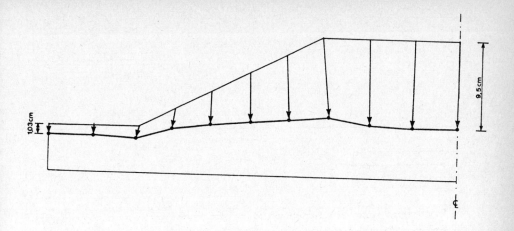

Figure 5.7.2 Upper Boundary Displacements.

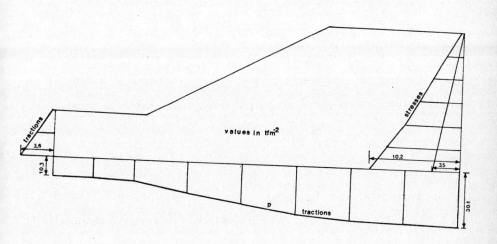

Figure 5.7.3 Tractions and Stresses.

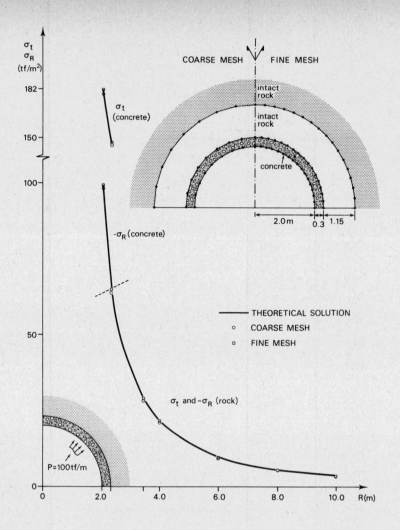

Figure 5.7.4 Lined Tunnel. Discretization and Results.

The problem has been solved using two different boundary meshes with 24 and 48 linear elements respectively. The elastic results (fig. 5.7.4) are in total agreement with the theoretical solution (100). The continuity and the accuracy of the stress profile on the two discretized interfaces show the validity of the subregion technique.

(c)   Circular tunnel in a prestressed rock medium

In many tunnel excavations in rock or soil materials, the assumption of plane conditions (generally plane strain) is readily justified. However, for cases in which the excavation direction does not coincide with a residual stress principal direction, out of plane displacements do not vanish and plane analysis cannot be applied. Instead of solving the problem as a three-dimensional case, the alternative is to assume complete plane strain conditions in which the final solution is computed as a combination of plane strain and anti-plane results.

The example chosen to illustrate this problem is a circular opening in a uniform prestressed rock (see fig. 5.7.5). The load applied is one due to the relief of initial compressive stress state on the real surface of the tunnel. Thus, constant radial and tangential tractions are applied all over the boundary.

The tunnel surface was discretized using 12 linear elements and 13 nodes. The symmetric conditions of the problem have been used, as shown by the discretization (fig. 5.7.6).

The plane and anti-plane displacements (fig. 5.7.7) obtained are compared with the theoretical solution (63) and they illustrate the good accuracy achieved, while by using finite element formulation less accuracy has been obtained for a similar tunnel (101).

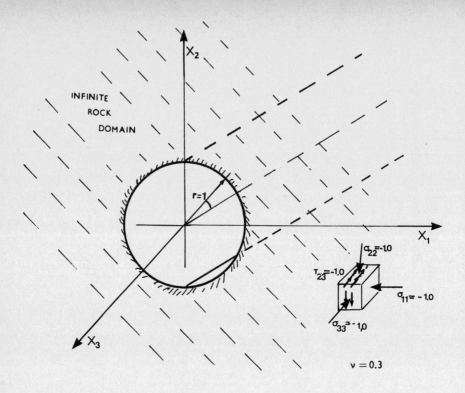

Figure 5.7.5 Circular Tunnel.

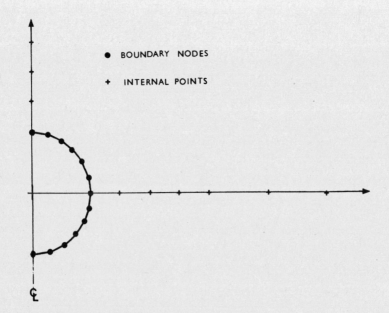

Figure 5.7.6 Tunnel Discretization.

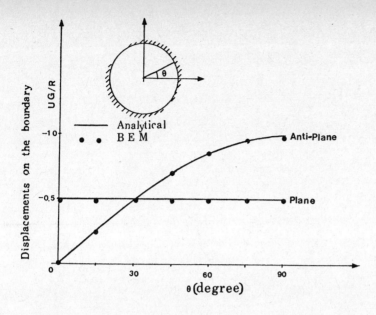

Figure 5.7.7 Displacements. Plane and Anti-Plane Cases.

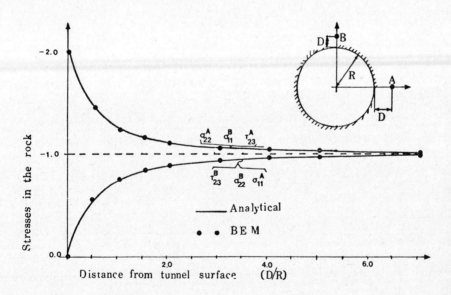

Figure 5.7.8. Stress Profile.

Figure (5.7.8) shows the stress distribution in the domain for plane and anti-plane problems and once more numerical results compare well with analytical solution.

(d)   Thin Elastic Seam Near a Circular Tunnel

This example is related to the application of the thin layer technique presented in section (5.5). The problem has been taken from reference (65) and consists of analysing a circular tunnel excavated between two seam regions inserted in a rock medium. The previous solution was obtained using a different boundary formulation in which the proximity of adjacent elements was taken into consideration by a particular type of singularity called quadrupoles.

Tunnel surface and seam interfaces are discretized into linear boundary elements as shown in figure (5.7.10). Then three subregions are defined; two of them are constituted of sound rock with Poisson's ratio taken equal to 0.3. The third subregion, the thin layer, is formed of soft soil to which two values of its elastic modulus have been assigned for the solution of the example, $E_s/E_R = 1.0$ and $E_s/E_R = 0.01$ respectively. As can be seen in figure (5.7.10), only one quadrant of the structural system was discretized. The load applied to the tunnel surface is one due to the removal of the initial uniaxial stress state ($\sigma_1 = -1.0$) which is assumed to exist before the excavation.

The final stress distribution achieved compares very well with theoretical (63) and numerical (65) results as shown in figure (5.7.11).

In spite of the necessary discretization of the thin layer, the formulation works with the same fundamental solution which has been used in this work. The final system of equations is similar to those obtained

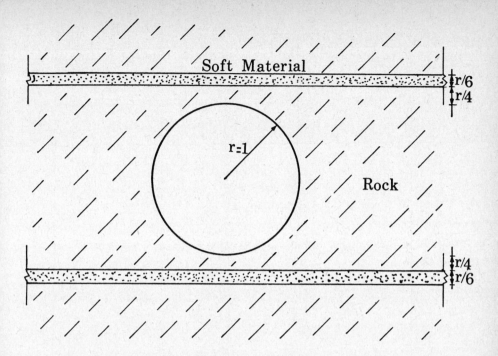

Figure 5.7.9 Thin Subregions in an Infinite Rock Medium.

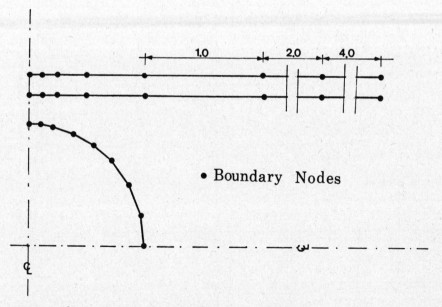

Figure 5.7.10 Boundary Discretization.

in problems involving subregions with banded matrices. The sizes of the seam elements do not need to be reduced according to the thickness of the layer. They have to be chosen according to the other elements in the adjacent subregions.

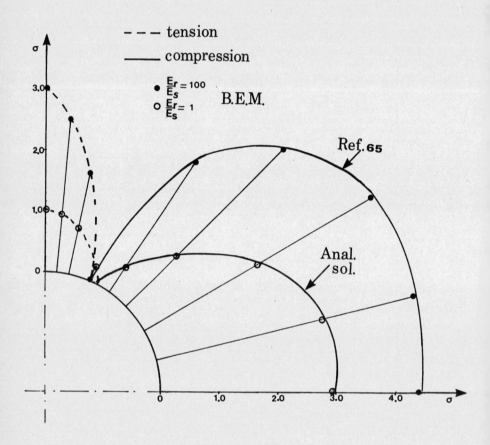

Figure 5.7.11 Normal Stress Distribution Parallel to the Tunnel Surface.

CHAPTER 6

NO-TENSION BOUNDARY ELEMENTS

6.1 Introduction

The boundary element equations presented in previous chapters are employed here to solve problems concerned with the theory of no-tension materials. The no-tension criterion appears to be one of the first departures from linear elastic theory used to model rock behaviour. In order to justify the choice of this criterion some aspects of rocks and rock masses are discussed, and the inability of these materials to sustain tensile stresses is emphasized in practical problems. The solution technique to be presented is achieved by an iterative process which consists of computing at each step an initial stress field to compensate the tensile stresses. The no-tension solution is shown to be obtained assuming a non-path dependent behaviour, therefore does not require any incremental process. An alternative solution considering path dependent behaviour is also presented.

Several examples are presented to illustrate the applicability of boundary elements to no-tension problems in geomechanics. The applications were selected to compare boundary element solution with either already published finite element results or theoretical solutions.

6.2 Rock Material Behaviour

The stress analysis for any rock excavation must be carried out bearing in mind the original state of the material usually assumed to be intact or fissured. For intact rock the analysis is performed within the context of continuum mechanics. In this case, the material does not show any joint or fissure and can be assumed homogeneous. Some rocks of the second group usually defined as rock masses are characterized by an excessive number of joints and fissures and can also be interpreted as a continuun.

Rock materials often present different responses when subjected to compressive or tensile stresses. Rock strength in tension is very small, and quite often rock mass shows to be incapable of sustaining any tensile stress, although when subjected to compressive stress the same mass can be assumed as a linear material. Thus, the no-tension criterion (46) appears to be a realistic assumption to model the material response. This material behaviour is provided by considering the rock as a linear elastic material in the direction of compressive principal stresses and simultaneously assuming no or very little resistance to deformations in the direction of principal tensions.

The no-tension analysis is specially appropriate to model stress redistribution around underground openings. In this case, the rock in the vicinity of opening surface cannot withstand any tensile stress due to either the original state of the fissured rock or the excavation process. The final stress state usually shows a free tension zone replacing the tensile region given by the linear elastic theory. The no-tension zones can also provide a good approximation for designing the temporary support required to prevent loosening over the roof.

The criterion is also suitable for a more accurate evaluation of the stresses in tunnels subjected to high internal pressure. Usually in this case, the rock surrounding the opening cannot sustain any tensile stresses due to the blasting process. The extent of the free tension zone is dependant upon the drilling process used if the rock material were initially intact. For tunnels built in rock mass, the inability of the material to sustain tensile stresses must be assumed everywhere and consequently all tension regions have to be eliminated.

6.3 Method of Solution

The no-tension criterion described in the previous section assumes that the material behaves differently for compressive or tensile stresses. In order to clarify the concept, the uniaxial relation between stress and strain for the no-tension case is presented in figures (6.3.1) and (6.3.2). In the first case (fig. 6.3.1), the problem is assumed not to be path dependent, and the complete recovery of the deformation when the load is removed is considered valid. Therefore, no incremental process is needed and the final solution can be computed applying the load in one step.

For the second uniaxial stress-strain relation (fig. 6.3.2), residual strains are expected on removal of the load and the final deformation state depends on both final stress level and path stress history. Thus, an incremental load process must be adopted to model the final solution.

The generalization to plane strain (or stress) situations is very simple. As the third coordinate axis is a principal direction, the corresponding stress is studied separately and does not affect the stresses on the plane. The only requirement needed to satisfy the criterion is that the plane principal stresses (σ_1 and σ_2) must remain less or equal to zero (fig. 6.3.3).

For the complete plane strain the Cartesian coordinate axes are not coincident with the stress principal directions, and all stresses

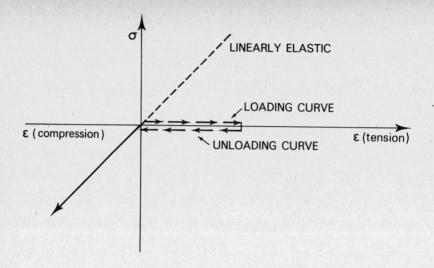

Figure 6.3.1 Uniaxial Stress-Strain Curve. Path Independent Criterion.

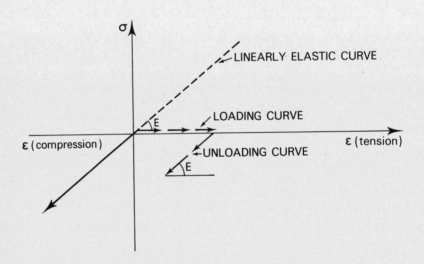

Figure 6.3.2 Uniaxial Stress-Strain Curve. Path Dependent Criterion.

must be taken into account together. Thus σ_1, σ_2 and σ_3 must assume non positive values to obey the no-tension condition (fig. 6.3.4).

The final solutions for the two cases described in figures (6.3.1) and (6.3.2) are modelled by iterative processes which consist of enforcing the no-tension condition (fig. 6.3.3 and 6.3.4) by computing an initial stress field (σ^o) at each step, employing equation (5.6.11).

Although the cases defined by figures (6.3.1) and (6.3.2) are obtained by similar process, in the non-path dependent criterion the principal stresses, which must obey the no-tension condition, are given by the equivalent elastic stresses $\underline{\sigma}^m$ (dashed line in fig. 6.3.1). For the path dependent criterion, the principal stresses are computed using the actual values of the stresses $\underline{\sigma}$ obtained after computing the corresponding initial stress field.

Bearing in mind the difference between the two criteria for the verification of the no-tension condition, the iterative process to model the final solution is illustrated in the following steps,

(i) Compute the elastic stress increment: from the elastic solution when the first iteration is performed, or from

$$\underline{\sigma}^e = \underline{S}\underline{\sigma}^o \qquad (6.3.1)$$

when the initial stress field $\underline{\sigma}^o$ has been applied. Notice that equation (6.3.1) represents equation (5.6.11) without vector $\underline{N}$ corresponding to the actual loads.

(ii) Find the stress vectors $\underline{\sigma}^m$ and $\underline{\sigma}$: vectors $\underline{\sigma}^m$ and $\underline{\sigma}$ are computed by the following expressions,

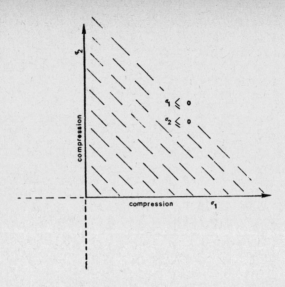

Figure 6.3.3 No-Tension Envelope for Two-Dimensional Problems.

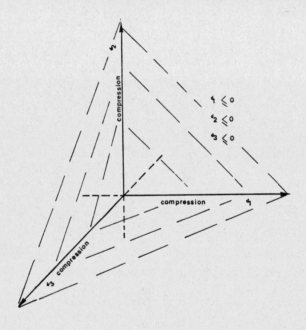

Figure 6.3.4 No-Tension Envelope for Complete Plane Strain and Three-Dimensional Problems.

$$\underset{\sim}{\sigma}^m + \underset{\sim}{\sigma}^e \to \underset{\sim}{\sigma}^m \qquad (6.3.2)$$

$$\underset{\sim}{\sigma}^t + \underset{\sim}{\sigma}^e \to \underset{\sim}{\sigma} \qquad (6.3.3)$$

in which $\underset{\sim}{\sigma}^t$ represents the true stresses obtained in the last iteration.

(iii) Compute the principal stresses : using $\underset{\sim}{\sigma}^m$ or $\underset{\sim}{\sigma}$ as dictated by path independent or path dependent material behaviour.

(iv) Determine the initial stress vector : the initial stress field to be applied to the system is computed by taking the values of $\underset{\sim}{\sigma}$ in the tensile principal direction obtained in (iii).

(v) Compute the true stress vector : the actual stress distribution is evaluated as follows,

$$\underset{\sim}{\sigma}^t = \underset{\sim}{\sigma} - \underset{\sim}{\sigma}^o \qquad (6.3.4)$$

(vi) Verify the convergence : at this stage the convergence criterion must be applied; if the initial stress vectors are small with respect to the specified tolerance, the iterative process should stop. Otherwise, all steps must be repeated.

It should be noticed that the vector $\underset{\sim}{\sigma}^m$ is needed only in the non-path dependent criterion. For the second criterion, $\underset{\sim}{\sigma}^m$ is neglected and the iterative process shown must be repeated for each increment of load.

The path dependent criterion can also be modelled considering a non-path dependent behaviour in each increment of load. The principal stresses are computed using $\underset{\sim}{\sigma}^m$ values which are forced to be equal to $\underset{\sim}{\sigma}^t$ when a new load increment is applied.

For the two cases presented, the displacements can be computed after applying all loads or, if necessary at any other stage, using the accumulated initial stress values in equation (5.6.4).

6.4 Application of No-Tension in Rock Mechanics

Four examples are presented here to illustrate the applicability of boundary elements in geomechanical no-tension problems. The first application is a steep valley already analysed using finite elements in reference (46). The second example consists of analysing a circular lined tunnel in which the region closer to the lining is considered as a no-tension zone. In the third application a deep tunnel also solved with finite elements (46) is analysed. The last application consists of analysing a semi-circular tunnel under action of a load applied on the ground surface.

(a) Steep Valley

This application consists of modelling the no-tension solution with the boundary element method for a steep valley previously solved by Valliappan (46). The load applied to the structural system is only due to the removal of the residual stresses which are assumed to be equal to $\bar{\gamma}y$ and $K_o\bar{\gamma}y$ in vertical and horizontal directions respectively (fig. 6.4.1). The elastic constants of the rock material to be excavated are: $G = 10^6$ lb/ft^2 and $\nu = 0.3$. The coefficient of earth pressure at-rest, K_o, is taken equal to 0.2, while the unit weight, $\bar{\gamma}$, is 150 lb/ft^2.

The sloped valley boundary is discretized into a series of straight elements over which tractions corresponding to the residual stresses are applied. In figure (6.4.2) the mesh used to solve the

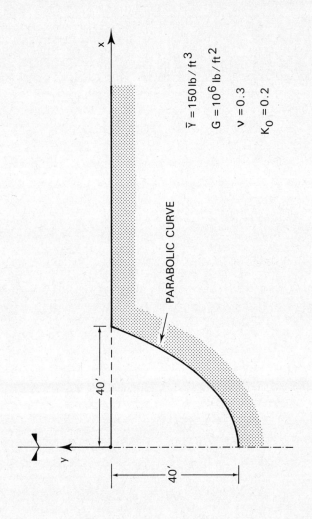

Figure 6.4.1 Steep Valley. Geometry.

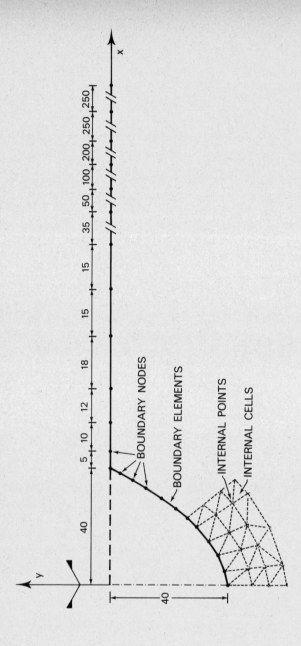

Figure 6.4.2 Steep Valley. Discretizations.

problem is presented. The accuracy of the results was verified using a finer mesh and practically the same results were obtained. For this problem the domain tends to infinity, therefore no displacement on the boundary has to be prescribed. Also the ground surface discretization was extended over a sufficient distance to simulate the infinite ground surface.

The two no-tension criteria have been used for the solution of this problem. Due to the type of load applied, the final solutions are very similar and compare well with the previous finite element results. For the first criterion, the initial elastic solution and the final no-tension solution are represented in the form of principal stresses plot as shown in figures (6.4.3) and (6.4.4) respectively. The figures also indicate the extent of the tension region for the elastic solution and the fissured zone in the final stress distribution.

It is important to notice that the boundary element formulation can represent the infinite domain accurately by discretizing the free surface into a small number of elements. The boundary solution required only 23 elements and 24 nodes on the boundary and 29 cells for the integration of the initial stress field in the domain, while the finite element solution was obtained using a mesh of 247 constant strain elements and 144 nodes.

(b) Circular Lined Tunnel

This example shows the analysis of a lined tunnel with internal pressure in an infinite rock medium. The only load applied in the system is an internal water pressure equal to 100 tf/m^2.

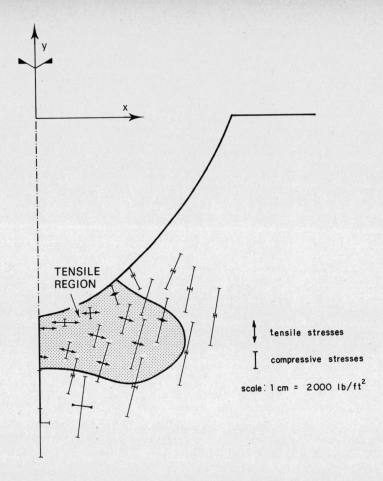

Figure 6.4.3 Elastic Stresses and the Tensile Zone.

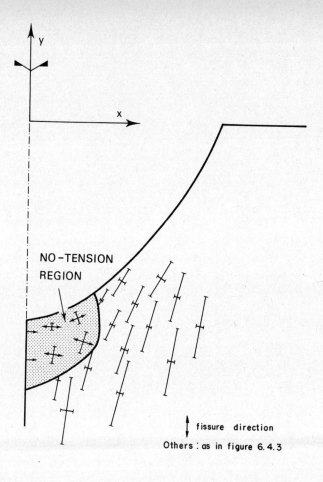

Figure 6.4.4 No-Tension Stresses and the Tension Free Zone.

The problem has already been defined in figure (5.9.4) in the last chapter, where the elastic solution was discussed. For the no-tension solution a 115 cm thick ring region closer to the lining is assumed fully fissured due to the blasting, and cannot withstand any tensile stress, while all other points outside that region are considered capable of sustaining any stress. The residual stresses in the rock were neglected in order to allow the comparison between numerical results and the theoretical solutions given in reference (102).

The problem was first solved by assuming the unlined case in which the load is applied directly to the rock material. Using 8 linear elements to discretize both the rock surface and the interface between intact and fissured material (fig. 6.4.5), the final no-tension results are obtained and compare well with the theoretical solution.

The no-tension analysis of the lined case has been carried out using two different internal cell discretizations and a single boundary mesh consisting of 8 linear elements and 9 nodes on the internal lining surface and on the interfaces (fig. 6.4.6). The final results for the coarsest mesh seem to be satisfactory; when they are compared with the theoretical solution (102), the maximum error observed was 4.5%. For the fine mesh, better agreement between theoretical and numerical solutions was obtained; the maximum computed error is only 2%.

(c) Bolted Tunnel

The tunnel example shown in fig. (6.4.7) and taken from reference (46) was used for a hydro-electric power station. The author analysed it as a no-tension problem with a finite element program using constant strain elements.

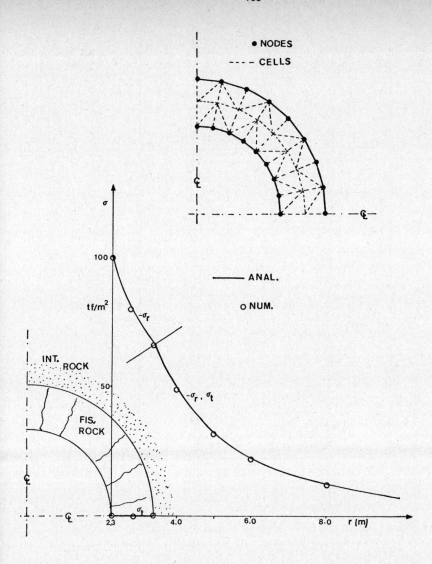

Figure 6.4.5 Unlined Case. Discretizations and No-Tension Results.

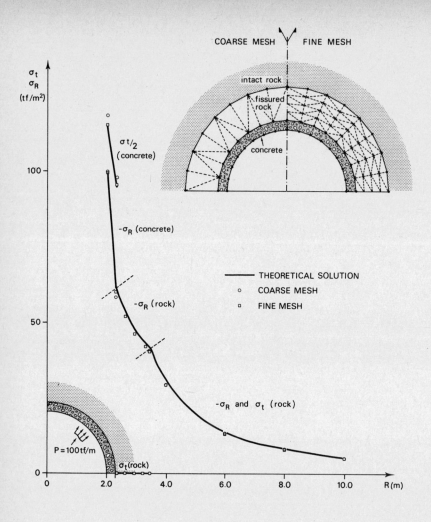

Figure 6.4.6 Lined Case. Discretizations and No-Tension Results.

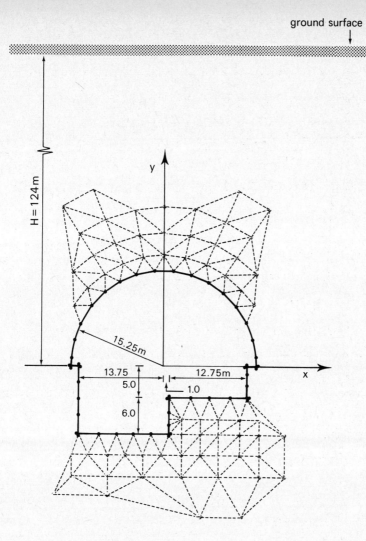

Figure 6.4.7 Discretizations. Tunnel Without Bolts.

The elastic modulus adopted for the rock is equal to 1.41×10^6 tf/m², while the Poisson's ratio is 0.15. The load applied is due to the relief of the residual stress state which is given by the unit weight $\bar{\gamma} = 2.5$ tf/m³ and $K_o = 0.2$.

The boundary mesh for the B.E.M. consists of 48 nodes and 39 linear elements. Two different internal discretizations have been employed to analyse the tunnel with and without prestress forces, as shown in figures (6.4.8) and (6.4.7) respectively.

The problem has been solved using the first no-tension criterion presented in section (6.3). The initial elastic and final no-tension solution for the analysis without bolting are presented in fig. (6.4.9) and they compare well with the finite element results. A large no-tension zone over the roof requires the use of structural lining or bolting to avoid rock falls. In this case, as shown in fig. (6.4.8), prestress forces radially applied are adopted. The final solution obtained (fig. 6.4.10) shows the reduction of the free tension zone in the vicinity of the tunnel boundary. New no-tension zones are now developed inside the domain.

(d) Semi-Circular Tunnel

The geometry of the semi-circular tunnel presented in fig. (6.4.11) was taken from reference (99). The type of the boundary discretization, the loading and the boundary conditions prescribed were chosen to make this example an optimum numerical test for the boundary technique rather than to reproduce an actual tunnel excavation problem.

The load consists of a large vertical distributed load P (fig. 6.4.11), which is applied on the ground surface. Residual stresses

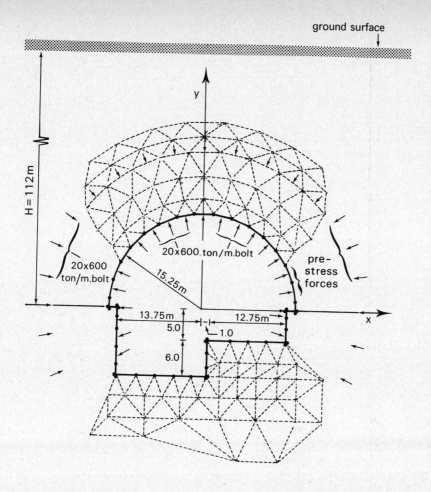

Figure 6.4.8 Discretizations. Tunnel with Bolts.

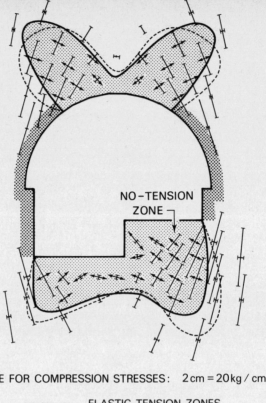

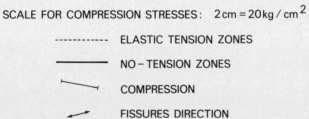

Figure 6.4.9   Tensile and No-Tension Regions. Tunnel Without Bolts.

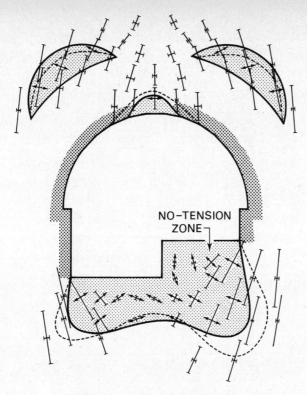

SCALE: 20kg/cm^2 for 1cm

-------------- ELASTIC TENSION ZONES
——————— NO TENSION ZONES
⊢———⊣ COMPRESSION
←—→ FISSURES DIRECTION

Figure 6.4.10 Tensile and No-Tension Regions. Tunnel With Bolts.

are assumed constant during the loading process and are computed according to both the unit weight of the material taken to be 120 lb/ft^3 and the maximum overburden, H , equal to 116 ft. The elastic constants assumed in this case are 2.16 × 10^7 tf/m^2 and 0.0, shear modulus and Poisson's ratio respectively.

The no-tension zones developed during the loading process are presented in fig. (6.4.12). The load was applied in 40 increments and the path dependent criterion was adopted. Figure (6.4.13) presents the displacements at points A and B to illustrate the nonlinear behaviour. The nonlinearities start at 4% of loading and continue up to 60% of it. At that stage of loading the final no-tension zone has already been reached and for all further displacements the structural system seems to behave as an equivalent linear elastic body.

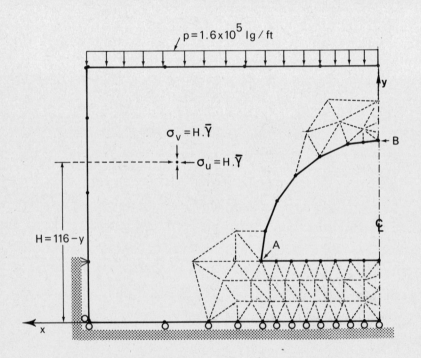

Figure 6.4.11 Semi-Circular Tunnel. Discretizations.

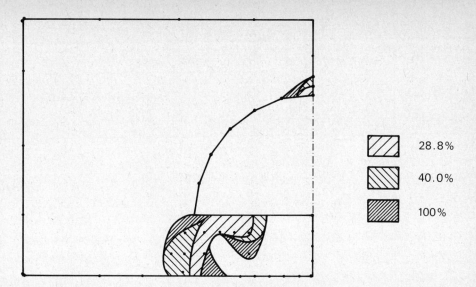

Figure 6.4.12 No-Tension Zones.

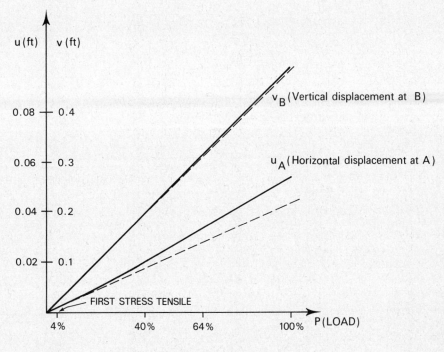

Figure 6.4.13 Displacements at "A" and "B".

CHAPTER 7

DISCONTINUITY PROBLEMS

7.1 Introduction

In this chapter alternative formulations to consider directionally orientated material weaknesses and slip or separation along discontinuities for rocks are presented. The chapter starts by introducing a model to study weakness in a particular direction within the context of continuum mechanics. In many schistous rock materials, shear and tensile strengths respectively parallel and orthogonal to schistosity are zero or very small in comparison to those of intact rocks (103). In order to model such material weaknesses the shear strength parallel to the schistosity is assumed to be governed by a simple Coulomb criterion of failure. Also the tensile stress in the direction orthogonal to the schistosity plane can be limited by the tensile strength value which is usually assumed to be zero.

The seam element formulation presented in Chapter 5 to model the thin layers often present in rock media is modified to simulate one or a series of rock discontinuities. A Coulomb criterion of failure is again assumed to govern the level of stress along the discontinuities. Equations to represent separation and slip between two rock blocks or subregions are developed. An iterative and incremental process is proposed to model the discontinuity solution. The technique adopted consists of modifying the main system of equations at each step according to the stress level along the discontinuities. Finally some examples are solved in order to illustrate the algorithm adopted.

7.2 Planes of Weakness

In schistous rock, the isotropic behaviour cannot be assumed valid. The elastic properties and the strength of the material have to be considered according to the direction of the schistosity. The elastic behaviour of such a kind of rock can be modelled by boundary element formulation since a convenient anisotropic fundamental solution (1) is used. The Kelvin fundamental solution can be used to model anisotropic elastic behaviour, but an inconvenient iterative process to correct the stresses to meet the real elastic properties must be employed.

The second important property shown by schistous rock is the considerable reduction of the shear strength parallel to the plane of schistosity (fig. 7.2.1). For plane problems, the stress tensor computed with reference to the local system of coordinates $(\bar{x}_1, \bar{x}_2)$ parallel to the schistocity must obey the Coulomb's law. The tension cut off in the direction orthogonal to the schistosity may be assumed either to eliminate or to reduce the tensile stresses in this direction.

For cases in which the plane of weakness or schistosity is not parallel to the third axis of coordinates, complete plane strain conditions can be assumed if no other argument is against it. Thus, the complete stress tensor with 9 components has to be used in order to determine the shear stresses on the weakness plane.

The plane of weakness model can also be adopted for highly fissured rock in which the fissures follow one preferential direction. In this case, the rock mass can be assumed to be an equivalent anisotropic material with reduced shear and tensile strengths parallel and orthogonal to the fissures plane respectively.

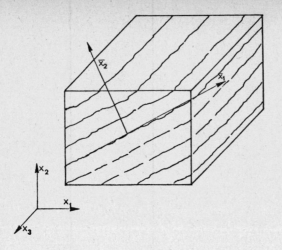

Figure 7.2.1 Schistous Rock Block.

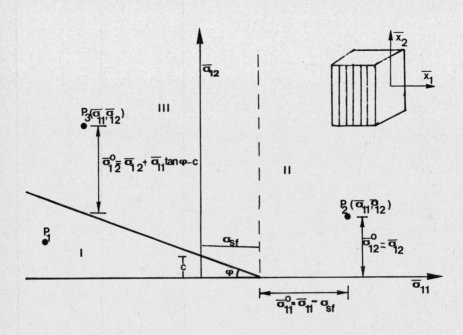

Figure 7.2.2 Assumed Envelope.

As for the no-tension solution, the conditions which must be met by the stress in the weakness direction are enforced by an iterative process consisting of applying initial stresses. In figure (7.2.2), the necessary initial stress field to correct the stresses of two points (P_2, P_3) outside region I are schematically shown, while no correction is needed for any point P_1 at which the stress components are limited by the following relations,

$$\bar{\sigma}_{11} \leqslant \sigma_{sf} \qquad (7.2.1)$$

$$\bar{\sigma}_{12} \leqslant - \bar{\sigma}_1 \tan\phi + c \qquad (7.2.2)$$

where σ_{st}, ϕ and c stand for the tensile strength, angle of friction and cohesion in the schistous directions.

The iterative process to model the final solution is given by the following steps:-

(i) Compute the elastic stress increment : from the elastic solution when the first iteration is performed; or from,

$$\underset{\sim}{\sigma}^e = \underset{\sim}{S} \underset{\sim}{\sigma}^o \qquad (7.2.3)$$

when the initial stress field $\underset{\sim}{\sigma}^o$ is applied.

(ii) Find the total stress vector : the vector $\underset{\sim}{\sigma}$ is computed by adding the elastic stress increment and the true stress vector $\underset{\sim}{\sigma}^t$ (at the beginning it is the residual stress vector) together, i.e.,

$$\underset{\sim}{\sigma} = \underset{\sim}{\sigma}^t + \underset{\sim}{\sigma}^e \qquad (7.2.4)$$

(iii) Determine the initial stress vector : writing the stress vector in a local system of coordinates, the initial stress vector $\underset{\sim}{\sigma}^o$ is obtained as schematically shown in figure (7.2.2.).

(iv) Compute the true stress vector : from $\underset{\sim}{\sigma}$ and $\underset{\sim}{\sigma}^o$ by,

$$\underset{\sim}{\sigma}^t = \underset{\sim}{\sigma} - \underset{\sim}{\sigma}^o \qquad (7.2.5)$$

(v) Apply the convergence criterion. If any $\underset{\sim}{\sigma}^o$ value is not zero with reference to the chosen tolerance, the process has to continue from "i". Otherwise, another increment of load can be applied.

This iterative process is highly dependent on the path of the stresses and requires incremental application of the load. A non-path dependent criterion which allows the load to be applied in only one increment is also possible, and is easily obtained using a vector $\underset{\sim}{\sigma}^m$ to control the elastic path as done for the no-tension criterion.

7.3 Analysis of Discontinuity Problems

Many types of rock are continuously disrupted by planar cracks, denoted fissures. When the rock material is under compressive state of stress, the cracks can be shut restoring the mechanical continuity of the material, while when under tension state of stress normal to the fissures, the separation increases and quickly destroys the existing cohesion. In this case, the rock material loses its continuity and must be analysed taking into account the possible separation or slip which may occur between the two surfaces defined by the discontinuity.

In order to model this behaviour using boundary element formulation the blocks of continuous material are assumed as subregions, while specific relations between the values of two opposite nodes along the discontinuity must be defined. The discontinuity can be interpreted as being formed by "joint element" with four nodes (fig. 7.3.1). The joint element width represents the real gap of the discontinuity which can be filled with soft material. The width is taken equal to zero when the subregions are in contact and the coordinates of the two opposite points are the same. The discontinuity has essentially no or limited resistance to a tension force applied in the normal direction. The slip between the two surfaces of the discontinuity is usually assumed to be governed by a linear Mohr envelope which may be also defined with limited tensile strength.

The boundary equations can be written for each of these nodes forming the joint element. In order to consider the joint elements, other conditions depending on the physical characteristics of the discontinuity must be also introduced to complete the final system of equations. These extra equations necessary to model any discontinuity problems are divided in three main groups. The first group represents the separation condition. This particular condition must be enforced to the problem when either a real gap between adjacent subregions exists or tensile stresses in the direction normal to the joint element exceed the admissible limit. The conditions to be enforced at each node "S" when no soft material is filling the gap can be expressed in the following form,

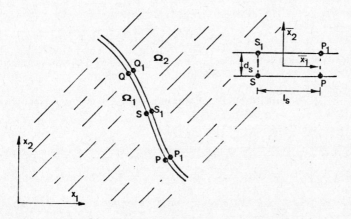

Figure 7.3.1 Discontinuity.

$$p_i^{(S)} = 0 \qquad (7.3.1)$$

in which $p_i^{(S)}$ represents interactive tractions between the two subregions. Any other load applied to the system at point "S" must be considered as an external traction.

Equation (7.3.1) can be easily introduced in the system of equations by considering each surface of the discontinuity as an external boundary. For cases in which the gap is considered filled, the contact still exists and the thin layer element relations replace equation (7.3.1). In this case, condition (7.3.1) is applied only when the separation between rock and soft material occurs.

The second group of equations necessary to model the discontinuity problem is formed by the contact equations. These equations are introduced to model the continuity condition between the two subregions, which is always associated with the stress level at the discontinuity. Considering a possible envelope for the stress condition at a discontinuity point as indicated in figure (7.3.2), the continuity is achieved only in region I. For points in this condition the mathematical expressions to be introduced into the system of equations are given by,

$$u_i^{(S)} - u_i^{(S_1)} = 0 \qquad (7.3.2)$$

$$p_i^{(S)} + p_i^{(S_1)} = 0 \qquad (7.3.3)$$

i.e., no relative displacement and no resultant of tractions are allowed between two opposite nodes on discontinuity surfaces.

The slip between the two surfaces of the discontinuity is the last condition to be formulated. This situation is represented by points in the region II of fig. (7.3.2). The shear strength has already been exceeded but no (or only limited) tensile stress occurs. Then, the two surfaces are supposed to slide on each other, without any relative displacement in the direction orthogonal to the discontinuity. In order to introduce such a behaviour in the system of equations, the following conditions of displacements and tractions in the direction orthogonal to the discontinuity can be written,

$$\bar{p}_2^{(S)} + \bar{p}_2^{(S_1)} = 0 \qquad (7.3.4)$$

$$\bar{u}_2^{(S)} - \bar{u}_2^{(S_1)} = 0 \qquad (7.3.5)$$

In the direction parallel to the discontinuity, equations relating the shear and normal stresses can be written according to the assumed Coulomb's envelope (fig. 7.3.2). Thus,

$$\bar{p}_1 = - \bar{p}_2 \tan\phi \qquad (7.3.6)$$

for positive values of the shear stress, otherwise

$$\bar{p}_1 = \bar{p}_2 \tan\phi \qquad (7.3.7)$$

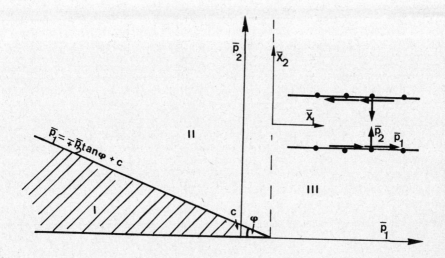

Figure 7.3.2 Discontinuity Conditions.

Equations (7.3.1) to (7.3.7) represent the new relations to be introduced into the system of equations. This can be done by changing the original boundary equations without creating new lines.

Discontinuity problems are markedly nonlinear; the stress level is not only function of the displacements, but also must be related to the history of the displacements. Therefore, the superposition is no longer valid and incremental procedure of loading must be followed to satisfy the path dependence of the problem.

The process of solution adopted in this work to deal with domains which show discontinuity consists of incremental and iterative procedure in which the system of equations may be modified at each iteration according to the stress values along the discontinuity.

The sequence of steps can be resumed as follows,

(i) Assemble the system matrix: the matrix $\underline{H}$ is computed taking into account the conditions defined at each node over the discontinuity

(ii) Apply boundary conditions: if it is the first iteration, the external boundary condition due to the load increment has to be applied. Otherwise traction computed in step "iii" is used.

(iii) Determine boundary traction to be applied along the discontinuity: if unadmissible tensile or shear stress were computed along the discontinuity, the traction vector necessary to reestablish the stress condition defined by the assumed envelope is computed, and the iterative process must start again at step "i". Otherwise, a new load increment has to be provided.

7.4 Numerical Applications

Three simple numerical applications are presented in this section to illustrate the applicability of the boundary element in problems involving discontinuities.

(a) Square Block

This example consists of analysing a square block with a possible discontinuity located at the middle section. The geometric characteristics and the boundary discretization of the block are given in fig. (7.4.1). Plane strain conditions have been assumed, and the Poisson's ratio and the elastic modulus are taken equal to 0.3 and 100000 kgf/cm^2 respectively. The shear stresses are governed by a Coulomb law characterised by the values of the cohesion $c = 2.0$ kgf/cm^2 and the friction angle $\phi = 0.0$.

Relative horizontal displacements between the two sides parallel to the discontinuity are enforced to be 0.0006 cm, while in the perpendicular direction the displacements are prescribed equal to zero. In the initial stage of the loading no crack forms and consequently the stress profile over the discontinuity surfaces is represented by a parabolic curve. For relative displacement equal to $\frac{1}{3}$ of the total prescribed value, the discontinuity has already occurred at the middle node at which the shear stress remains equal to the limit governed by the cohesion c as can be seen in figure (7.4.2); at that same stage of load, the continuity of the block is observed at other nodes, as is shown in fig. (7.4.3), in which horizontal displacements of a vertical side are plotted. When the total displacement is enforced, the shear stresses at all discontinuity nodes reach the limit c (see fig. 7.4.2) and

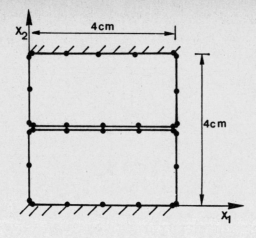

Figure 7.4.1 Square Block. Geometry and Discretization.

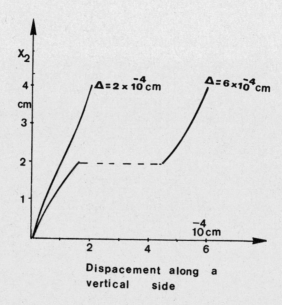

Figure 7.4.2 Displacements.

relative displacements between the two surfaces of the discontinuity take place, as shown in figure (7.4.3).

(b)   Unsupported Weight of Soil

Another example solved by assuming slip or separation along one defined surface is the case of a vertical soil cut. The example consists of analysing a vertical cut of height equal to 8.66 m as is shown in figure (7.4.4). A straight steep surface from the cut bottom to a point on the original ground 5m horizontally distant is assumed to be a weakness plane with the shear stress governed by the Coulomb's friction law. The cohesion c and the friction angle ϕ are taken equal to $0.2 kgf/cm^2$ and $30°$ respectively. The discretization used for the boundary element solution is given in figure (7.4.4), in which the slip plane discretized as a discontinuity surface makes an angle β equal to $60°$ with the horizontal direction. The load due to the unit weight, $\bar{\gamma}$, equal to $2.0 \, tf/m^2$ is applied incrementally, enforcing plane strain conditions. Figure (7.4.5) presents the displacement at node A against the load factor (percentage of the unit weight applied). These results show that the maximum load factor before the collapse of the system is equal to 0.8, i.e., the applied load corresponds to a unit weight $\bar{\gamma} = 1.6 \, kgf/cm^2$ which compares well with the analytical results obtained assuming the mechanism of rigid blocks (ref. 53). Figure (7.4.6) shows the shear and normal stress distributions along the discontinuity surface; at all nodes the shear stresses have practically reached their maximum values according to the Coulomb law. Figure (7.4.7) shows the final displacement for the external nodes; as the load resultant applied to the system is not zero, only the relative displacements have physical meaning. All absolute displacement values in the "X_2" direction are associated with rigid body translation governed by the Kelvin fundamental solution.

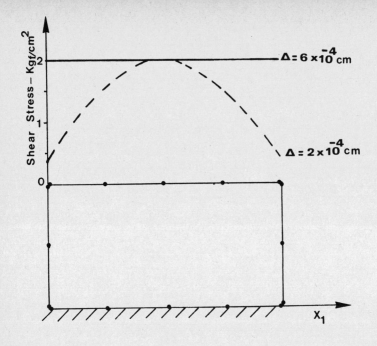

Figure 7.4.3 Shear Stress Distribution Along the Discontinuity.

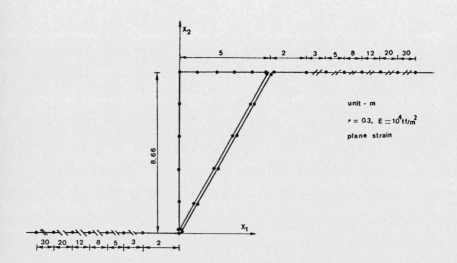

Figure 7.4.4 Unsupported Height of Soil.

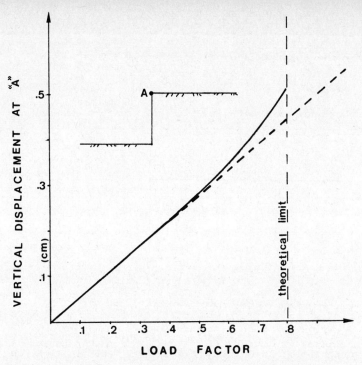

Figure 7.4.5 Vertical Displacements at "A".

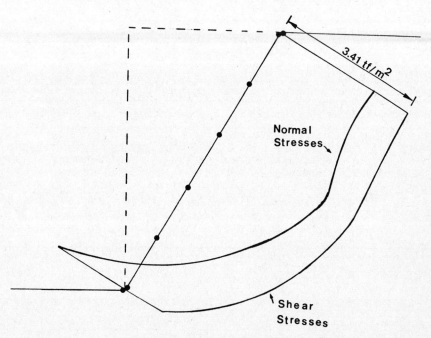

Figure 7.4.6 Stress Distributions Along the Discontinuity.

(c) Circular Opening in a Cracked Rock Domain

The application consists of solving a circular tunnel built in a solid rock medium with two joints parallel to the opening axis. The Poisson's ratio assumed for the rock material is 0.3. On the discontinuity the shear stresses are governed by a simple Coulomb's friction law for which the friction angle ϕ and the cohesion c are assumed to be equal to 30° and 0.0 respectively. The discretizations for the boundary and crack are presented in fig. (7.4.8); as can be seen, no relative displacement between the two discontinuity surfaces exists before the application of the load. Only horizontal loads are applied to the system in order to relieve the uniaxial system of residual compressive horizontal stresses ($\sigma_{11} = -1.0$). The results obtained with the boundary technique are shown in figures (7.4.9) and (7.4.10). In the first figure the new position of the boundary nodes is presented in comparison with the corresponding displacements for non fissured case. The second diagram shows the distribution of the normal stress in the tangential direction over the tunnel boundary. These results also illustrate the effects of the discontinuity when compared with the solution obtained using intact rock medium.

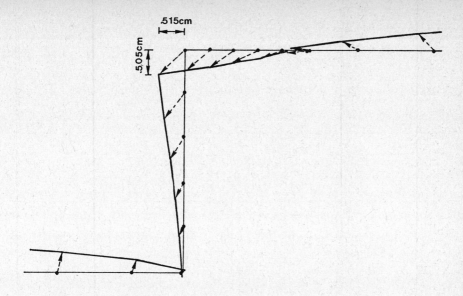

Figure 7.4.7 Final Boundary Displacements.

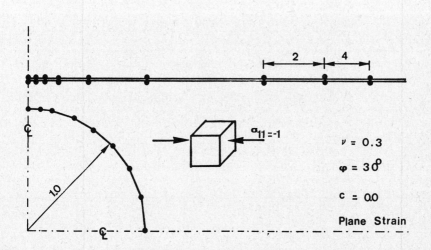

Figure 7.4.8 Circular Tunnel and Discontinuity Surface. Geometry and Discretization.

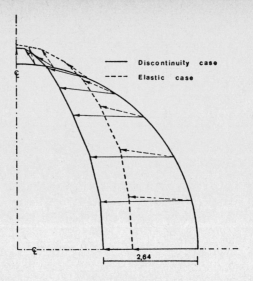

Figure 7.4.9 Boundary Displacements.

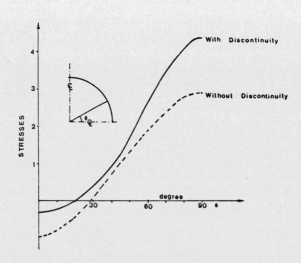

Figure 7.4.10 Normal tangential Stress on the Tunnel Surface.

CHAPTER 8

BOUNDARY ELEMENT TECHNIQUE FOR PLASTICITY PROBLEMS

8.1 Introduction

In this chapter the boundary element formulation presented so far is adapted to solve problems concerned with the classical theory of plasticity. The chapter starts by reviewing the essential features of the one-dimensional elastoplastic analysis, followed by generalization of the plasticity concepts to continuum problems. Stress-strain relationships for post yield conditions are formulated for the boundary element technique. The procedure to compute the plastic solution is based on the initial stress process proposed by Zienkiewicz (42) for finite element formulation. The procedure has been implemented to handle four well established yield criteria (Mohr-Coulomb, Drucker-Prager, von Mises and Tresca).

Several applications mainly related to geomechanical problems are presented to illustrate the usefulness of the technique. Most of the problems solved are related to underground excavation particularly suitable for boundary formulation due to the infinite domains involved.

8.2 Elastoplastic Problems in One Dimension

Elastoplastic behaviour is another possible departure from the elastic theory. The material is assumed to behave elastically only in a certain range of small strains characterized by an elastic limit. After reaching this limit, complete recovery of the original unstrained state of the material through the unloading process is no longer possible and the strain configuration is dependent on both stress level and load history. Another characteristic also associated with the plasticity theory is its total independence of time. The plastic deformations are assumed to be developed instantaneously as the load is applied.

For one-dimensional situations the elastoplastic behaviour can be defined by the parameters obtained from the uniaxial test. In order to simplify the plastic analysis some idealized stress-strain curves are assumed to represent the behaviour of the material.

The simplest idealized elastoplastic behaviour is given by figure (8.2.1) in which identical response is assumed in tension and in compression. This behaviour, which is adopted for elasto perfectly plastic material, is also represented by a rheological model consisting of a Hookean spring and a slider in series (fig. 8.2.2). The behaviour is purely elastic until the stress level reaches the yield stress, when the slider is supposed to yield and the deformation increases indefinitely. Considering that the point A in figure (8.2.1) represents the stress condition of a loaded body, the unloading curve is given by the dashed line parallel to the elastic path. Therefore, an irreversible deformation, ε^p, is left in the body after the removal of the load. For any other applied load (whether tension or compression) the body behaves elastically before reaching again the yield stress, σ_y, and the total deformation must include the plastic strain, ε^p, due to the previous loading.

The total deformation due to any load applied can be separated into elastic, ε^e, and plastic, ε^p, components, i.e.,

$$\varepsilon = \varepsilon^e + \varepsilon^p \qquad (8.2.1)$$

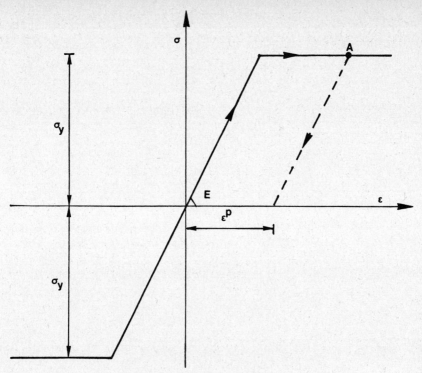

Figure 8.2.1 Uniaxial Stress-Strain Curve. Elastic Perfectly Plastic Material.

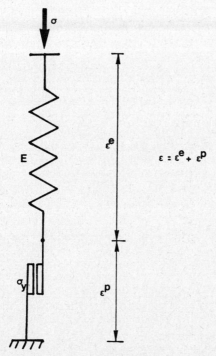

Figure 8.2.2 Rehological Model. Elastic Perfectly Plastic Material.

The stress level for an elasto perfectly plastic material must obey the following condition,

$$\sigma - \sigma_y \leq 0 \qquad (8.2.2)$$

in which only positive values of σ have been considered, due to the symmetry of the stress-strain relations.

The left hand side of (8.2.2) can be interpreted as a yield function $F(\sigma)$ which cannot assume positive value, i.e,

$$F(\sigma) \leq 0 \qquad (8.2.3)$$

Another idealised elastoplastic behaviour consists of assuming hardening (or softening) effects developed after the yield limit. Figure (8.2.3) shows a simplified stress-strain relation with linear hardening characterized by the constant modulus E_T. In this case, the behaviour can be also represented by a rheological model figure (8.2.4). Initially, the elastic responses are given by two springs in parallel. When the stress level reaches the yield limit, σ_y, the slider yields and only one spring will work to model the linear hardening response. Under these conditions, the yield stress does not represent the maximum stress level possible; it only signifies the beginning of the plastic flow. If the point A in figure (8.2.4) indicates the stress-strain level for a loaded body, the unloading curve is also, in this case, given by the line parallel to the elastic path. In subsequent loadings corresponding to either positive or negative values of σ, the material will yield again only when the stress level reaches the value Y, and the previous deformation must be taken into account for computing the new value of the plastic strain.

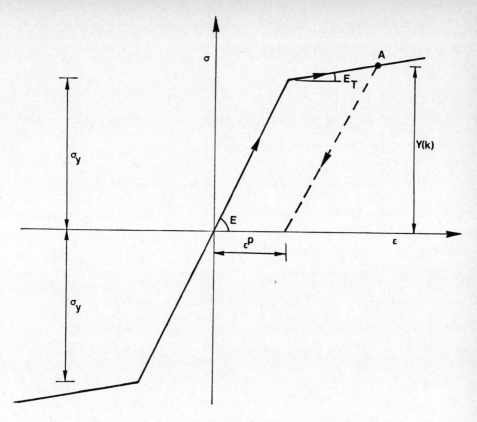

Figure 8.2.3 Uniaxial Stress-Strain Curve. Elastoplastic Behaviour with Hardening Effects.

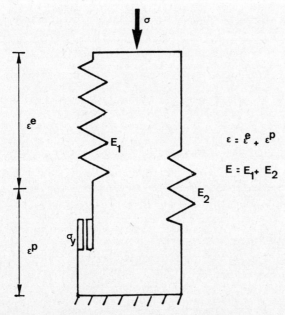

Figure 8.2.4 Rheological Model. Elastoplastic Behaviour With Hardening Effects.

Equation (8.2.1) remains valid as a representation of the total strain, and the stress level is governed by the following condition,

$$F(\sigma, k) \leq 0 \qquad (8.2.4)$$

or

$$\sigma - Y(k) \leq 0 \qquad (8.2.5)$$

in which k represents the hardening (or softening) parameter.

In plasticity theory hardening effects are often taken into account by applying the work hardening concept. Using this concept the yield stress can be interpreted as a function of the hardening parameter k, which is associated with the total plastic work done per unit of volume, i.e.,

$$k = \int \sigma d\varepsilon^p \qquad (8.2.6)$$

or

$$Y = Y(\int \sigma d\varepsilon^p) \qquad (8.2.7)$$

For linear hardening (represented by the constant slope in figure 8.2.3) the yield stress is computed by,

$$Y = Y(k) = \sigma_y + H'\varepsilon^p \qquad (8.2.8)$$

in which

$$H' = E_T/(1-E_T/E) \qquad (8.2.9)$$

The plastic behaviour discussed so far deals only with isotropic plastic material in which the symmetry of the stress-strain curve is valid. However, after the first plastic deformation, many materials can lose the symmetry property and the yield stress levels for tension and compression can not be represented by the same value $Y(k)$. A perhaps better explanation of this effect is one based on the anisotropy of the dislocation field produced by loading. This effect is known as the Bauschinger effect and is present whenever there is a reversal of the stress field. In order to take into consideration a kind of Bauschinger effect, a combination of several stress-strain relations, each one associated with a weighted parameter as proposed by Besseling (116) and later implemented for finite element formulation (47-49), can easily be adopted. This procedure, known as the "overlay technique", will be presented and introduced into boundary element methods in the following chapter to deal with more complex time-dependent effects.

8.3 Theory of Plasticity for Continuum Problems

In the previous section the plastic behaviour for one-dimensional problems has been presented. Its generalization to deal with continuum problems is summarized here together with the necessary basic assumptions.

In continuum mechanics, the total strain, ε_{ij}, can be also expressed as the sum of the elastic, ε_{ij}^e, and plastic, ε_{ij}^p components i.e.,

$$\varepsilon_{ij} = \varepsilon_{ij}^e + \varepsilon_{ij}^p \qquad (8.3.1)$$

and similarly a small increment of strain $d\varepsilon_{ij}$ can be decomposed into elastic and plastic terms as follows,

$$d\varepsilon_{ij} = d\varepsilon_{ij}^e + d\varepsilon_{ij}^p \qquad (8.3.2)$$

In order to formulate a theory which models elastoplastic material responses, the following general relations between stress and strain must be specified,

(a) Explicit elastic stress-strain relationship.

(b) Yield criterion which indicates the onset of plastic flow and an end of the elastic behaviour.

(c) Relationship between stress and strain for post yield behaviour in order to compute the plastic strain.

The elastic stress-strain relation to meet the first requirement is given by equation (3.2.12), in which the initial strain term $\varepsilon_{k\ell}^o$, must be neglected if no temperature or similar load is prescribed, i.e.,

$$\sigma_{ij} = C_{ijk\ell}\varepsilon_{k\ell}^e \qquad (8.3.3)$$

The yield criterion for determining the stress level at which plastic deformations begin can be written as,

$$F(\sigma_{ij}, k) = 0 \qquad (8.3.4)$$

or

$$f(\sigma_{ij}) - Y(k) = 0 \qquad (8.3.5)$$

in which $f(\sigma_{ij})$ is a function only of the stress values and can be interpreted as an equivalent stress value σ_e for a one-dimensional problem; and $Y(k)$ given as a function of the hardening parameter k plays the role of the yield stress value.

Any yield criterion for an isotropic material must be independent of the system of coordinates employed, therefore it should be a function of the three invariants only, which are given by,

$$\begin{aligned} I_1 &= \sigma_{kk} \\ J_2 &= \frac{1}{2} S_{ij} S_{ij} \\ J_3 &= \frac{1}{3} S_{ij} S_{jk} S_{ki} \end{aligned} \qquad (8.3.6)$$

in which the deviatoric stresses, S_{ij}, are defined by

$$S_{ij} = \sigma_{ij} - \delta_{ij}\sigma_{kk}/3 \tag{8.3.7}$$

An alternative representation for these invariants is given by the angular form of the third invariant originally introduced by Lode (117). The invariant now is denoted by θ_o and is given by,

$$\theta_o = \frac{1}{3}\sin^{-1}\left[-\frac{3\sqrt{3}}{2}\frac{J_3}{J_2^{3/2}}\right] \tag{8.3.8}$$

in which the following condition must be obeyed,

$$-\frac{\pi}{6} \leq \theta_o \leq \frac{\pi}{6} \tag{8.3.9}$$

Now using the three invariants I_1, J_2 and J_3 or θ_o we can easily define various yield criteria in the form given by equation (8.3.4). For soil and rock material analysis the Mohr-Coulomb and Drucker-Prager criteria are the usual representation to separate purely elastic stress from plastic stress states. The first is represented in three-dimensional principal stress by angular pyramidal surfaces, while the other is represented by right circular cones.

The Mohr-Coulomb yield criterion is obtained by a generalization of the Coulomb friction failure law (106) defined by,

$$\tau = c + \sigma_n \tan\phi \tag{8.3.10}$$

where c is the cohesion, σ_n the normal stress on the failure plane, ϕ the angle of internal friction and τ the magnitude of the shearing stress.

This law can be represented by a straight line tangent to the largest Mohr circle, as shown in figure (8.3.1), in which the tensile stresses are assumed to be positive, and $\sigma_1 \leq \sigma_2 \leq \sigma_3$.

Enforcing the limit of the shear stresses represented in figure (8.3.1), a final yield condition involving the stresses (represented by the invariants), the cohesion value c, and the internal friction angle ϕ is derived as follows,

$$F(\sigma_{ij},c) = \frac{I_1}{3} \sin\phi + \sqrt{J_2} \left(\cos\theta_o - \frac{1}{\sqrt{3}} \sin\theta_o \sin\phi\right) - c \cos\phi = 0 \qquad (8.3.11)$$

When the angle of friction ϕ is taken equal to zero, the Mohr-Coulomb yield function reduces to,

$$F(\sigma_{ij},Y(k)) = \sqrt{J_2} \cos\theta_o - \frac{Y(k)}{2} = 0 \qquad (8.3.12)$$

in which the cohesion c is substituted by the value $Y(k)/2$ taken from the one-dimensional representation.

Equation (8.3.12) represents the surface for the Tresca yield criterion of maximum shear stress. This criterion, more suitable for metal applications, can also be used in non frictional soil stress analysis.

The other yield criterion commonly used in geomechanical applications is due to Drucker and Prager (12). This criterion has a smooth surface representation in the principal stress space and can be interpreted as a generalization of the Coulomb rule now written as a function of the two first invariants, I_1 and J_2. The yield function, in this case, was formulated modifying the von Mises expression by introducing the influence of the hydrostatic stress component which is represented by the first invariant, I_1. Thus, the final yield condition can be expressed by,

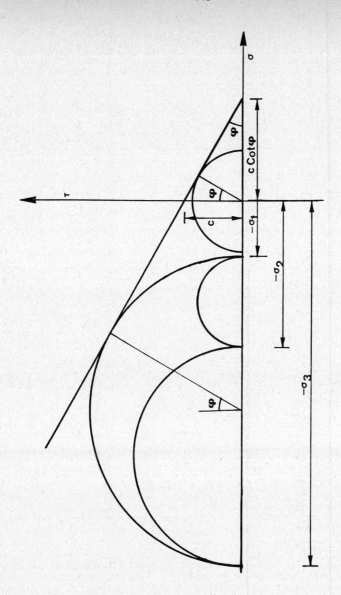

Figure 8.3.1 Mohr Circle Representation of the Mohr-Coulomb Yield Criterion.

$$\alpha I_1 + J_2^{\frac{1}{2}} - K = 0 \qquad (8.3.13)$$

with the constant α and K defined in terms of the usual material parameters, cohesion c and angle of friction ϕ, while the original von Mises' surface is given by,

$$J_2^{\frac{1}{2}} - \frac{Y(k)}{\sqrt{3}} = 0 \qquad (8.3.14)$$

Many α and K values have already been suggested in several works in order to achieve a better representation of plastic behaviour for soil and rocks. The original values given in reference (12) try to reproduce the Mohr-Coulomb hypothesis for plane strain conditions. In this case, the constants of equation (8.3.13) are given by the following expressions,

$$\alpha = \frac{\tan\phi}{(9+12\tan^2\phi)^{\frac{1}{2}}} \quad , \quad K = \frac{3c}{(9+12\tan^2\phi)^{\frac{1}{2}}} \qquad (8.3.15)$$

Another approximation of this type can be obtained enforcing the Drucker-Prager circle to be coincident with the outer apices of the Mohr-Coulomb hexagonal surface at any section. Then α and K are given by,

$$\alpha = \frac{2\sin\phi}{3(3-\sin\phi)} \quad , \quad K = \frac{6c\cos\phi}{3(3-\sin\phi)} \qquad (8.3.16)$$

If the coincidence with the inner apices of the Mohr-Coulomb hexagonal surface were enforced, the values α and K would become,

$$\alpha = \frac{2\sin\phi}{\sqrt{3}(3+\sin\phi)}, \quad K = \frac{6c\cos\phi}{\sqrt{3}(3+\sin\phi)} \qquad (8.3.17)$$

Another assumption which must be made in order to formulate plastic material behaviour is related to the determination of the plastic strain. This value is assumed to be uniquely determined from the state of stress by using the gradient of the plastic potential Q, which is a scalar function of both stress state and hardening parameter. Thus, any plastic strain increment is determined by,

$$d\varepsilon_{ij}^{P} = d\lambda \frac{\partial}{\partial \sigma_{ij}} G(\sigma_{ij},k) \qquad (8.3.18)$$

in which $d\lambda$ is a proportionality constant termed the plastic multiplier.

If the plastic potential $G(\sigma_{ij},k)$ is chosen to be represented by the same yield function $F(\sigma_{ij},k)$ which governs the elastic limit, the flow rule (eq. 8.3.18) is said to be associative. For cases in which $G(\sigma_{ij},k)$ has a more general representation independent of $F(\sigma_{ij},k)$, a non associative flow rule is defined.

In order to derive the stress-strain relationship for post yield behaviour for the general case, a procedure given in reference (39) has been followed. The formulation starts by writing the stress increment value substituting the elastic strain increment (eq. 8.3.2) in equation (8.3.3). Thus,

$$d\sigma_{ij} = C_{ijk\ell}(d\varepsilon_{k\ell} - d\varepsilon_{k\ell}^{P}) \qquad (8.3.19)$$

Introducing the plastic strain increment calculated by the flow rule (8.3.18) gives,

$$d\sigma_{ij} = C_{ijk\ell}(d\varepsilon_{k\ell} - a^{(G)}_{k\ell}d\lambda) \qquad (8.3.20)$$

in which

$$a^{(G)}_{k\ell} = \frac{\partial G}{\partial \sigma_{k\ell}}(\sigma_{k\ell},k) \qquad (8.3.21)$$

Considering that the derivative of the yield function $F(\sigma_{ij},k)$ (eq. 8.3.4) is also zero one can write,

$$a^{(F)}_{ij} d\sigma_{ij} - \frac{\partial Y(k)}{\partial k} dk = 0 \qquad (8.3.22)$$

in which

$$a^{(F)}_{ij} = \frac{\partial}{\partial \sigma_{ij}} F(\sigma_{ij},k) = \frac{\partial}{\partial \sigma_{ij}} f(\sigma_{ij}) \qquad (8.2.23)$$

Introducing the work hardening concept (eq. 8.2.6) now for a multiple state of stress, one has,

$$k = \int \sigma_{ij} d\varepsilon^p_{ij} \qquad (8.3.24)$$

which can be substituted in (8.3.22) to give,

$$a^{(F)}_{ij} d\sigma_{ij} - \frac{\partial Y(k)}{\partial k} \sigma_{ij} a^{(G)}_{ij} d\lambda = 0 \qquad (8.3.25)$$

Taking into account that the function $G(\sigma_{ij},k)$ can also be represented by,

$$G(\sigma_{ij},k) = g(\sigma_{ij}) - Y(k) \qquad (8.3.26)$$

the second term of equation (8.3.25) can be written as

$$\frac{\partial Y(k)}{\partial k} \sigma_{ij} a^{(G)}_{ij} d\lambda = \frac{\partial Y(k)}{\partial k} \sigma_{ij} \frac{dg}{d\sigma_{ij}}(\sigma_{ij}) d\lambda \qquad (8.3.27)$$

which, after applying the Euler's theorem, becomes,

$$\frac{\partial Y(k)}{\partial k} \sigma_{ij} a_{ij}^{(G)} d\lambda = \frac{\partial Y(k)}{\partial k} g(\sigma_{ij}) d\lambda \qquad (8.3.28)$$

Interpreting $g(\sigma_{ij})$ as an equivalent one-dimensional stress σ_e and using the work hardening definition (eqs. 8.2.6 and 8.3.24), we can write again equation (8.3.25) in the following form,

$$a_{ij}^{(F)} d\sigma_{ij} - H' d\lambda = 0 \qquad (8.3.29)$$

in which H' is a function of the uniaxial slope, as shown in equation (8.2.9).

Using equations (8.3.20) and (8.3.29), the proportionality parameter is derived as follows,

$$d\lambda = \frac{a_{ij}^{(F)} C_{ijk\ell} d\varepsilon_{k\ell}}{\left[a_{ij}^{(F)} d_{ij} + H'\right]} \qquad (8.3.30)$$

in which

$$d_{ij} = C_{ijk\ell} a_{k\ell}^{(G)} \qquad (8.3.31)$$

Replacing the $d\lambda$ value in equation (8.3.20) gives,

$$d\sigma_{ij} = \left[C_{ijk\ell} - \frac{d_{ij} a_{mn}^{(F)} C_{mnk\ell}}{a_{mn}^{(F)} d_{mn} + H'} \right] d\varepsilon_{k\ell} \qquad (8.3.32)$$

The above expression is the relation between the stress and strain increments for post-field behaviour.

For computational purposes, it is convenient to separate the stress increment into elastic and plastic components, i.e.,

$$d\sigma_{ij} = d\sigma_{ij}^e - d\sigma_{ij}^p \qquad (8.3.33)$$

with the elastic increment given by,

$$d\sigma_{ij}^e = C_{ijk\ell} d\varepsilon_k \qquad (8.3.34)$$

while the plastic term is computed by,

$$d\sigma_{ij}^p = \frac{d_{ij} a_{mn}^{(F)}}{\left[a_{mn}^{(F)} d_{mn} + H' \right]} d\sigma_{mn}^e \qquad (8.3.35)$$

It is important to notice that all tensor values used in this section are referred to three-dimensional representation. So all subscript indices have a range of three. For the plane case some simplifications to the tensors $a_{mn}^{(F)}$ and d_{mn} are needed. These simplifications can be seen in Appendix B where vector representation of these tensors are presented for plane stress, plane strain and complete plane strain cases.

The Drucker-Prager and von Mises yield surfaces presented in this section are smooth surfaces in the principal stress space, consequentl their derivatives with reference to the stress tensor components at every point are uniquely defined. On the other hand, Mohr-Coulomb and Tresca surfaces have corners located by $\theta_o = 30°$ resulting in the indetermination of the plastic strain. In this work, the procedure adopted to overcome these difficulties is simply to round the corners of the surface when it is approached within a certain tolerance. In order to achieve this, the derivatives of the Mohr-Coulomb and Tresca surfaces are substituted by equivalent values obtained using von Mises and Drucker-Prager respectively when $|\theta_o| > 29°$; i.e., the tolerance adopted is $1°$.

8.4 Numerical Approach for the Plastic Solution

In the previous section we have shown the basic requirements necessary to model plastic solutions. Now using the boundary element formulation, a numerical approach similar to the intial stress process given in reference (42) for finite elements is presented.

Due to the nature of equation (8.3.32), which expresses an incremental relation between stress and strain for post yield conditions, the numerical process for the plastic solution requires that the loads are applied in increments. For a generic increment of load, the problem is solved elastically and the elastic increment ($d\sigma_{ij}^e$) is added to the actual stresses. If any point reaches the plastic stage, the actual stress increment has to be computed and excedent stresses or plastic stress increments ($d\sigma_{ij}^p$) have to be applied to the system as an initial stress field (σ_{ij}^o).

The usual sequence adopted for each increment of load can be summarized by the following steps,

(i) Compute elastic stress increment vector $\Delta\underset{\sim}{\sigma}^e$. It is given by vector $\underset{\sim}{N}$ (eq. 5.8.11) when the load increment is being considered (first iteration of each increment) or by,

$$\Delta\underset{\sim}{\sigma}^e = S\Delta\underset{\sim}{\sigma}^p \qquad (8.4.1)$$

when the exceeding stresses ($\underset{\sim}{\sigma}^o = \Delta\underset{\sim}{\sigma}^p$) are applied.

(ii) Determine plastic stress increment and actual (or true) stress increment vectors ($\Delta\underset{\sim}{\sigma}^p$, $\Delta\underset{\sim}{\sigma}$). By using equations (8.3.35) and (8.3.33) respectively.

(iii) Compute the actual stress vector and the accumulated initial stress values. These values are obtained by,

$$\underset{\sim}{\sigma} + \underset{\sim}{\Delta\sigma} \to \underset{\sim}{\sigma} \qquad (8.4.2)$$

$$\underset{\sim}{\sigma}^a + \underset{\sim}{\sigma}^o \to \underset{\sim}{\sigma}^a \qquad (8.4.3)$$

(iv)   Verify the convergence.

If the plastic strain increment values are too small in comparison with the limit value given by the yield criterion function, the iterative process for the load increment should stop.

(v)   Go to step "i" for either another iteration or a new load increment.

In the approach shown above, only the stress values are computed, i.e., only equation (5.6.11) has been used. The boundary value vector $\underset{\sim}{X}$, given by equation (5.6.4) is not referred to in the iterative process, but its components can be computed at any time using either the plastic stress increment $\Delta\underset{\sim}{\sigma}^p$ or its accumulated value $\underset{\sim}{\sigma}^a$. For a problem in which only final displacement and traction profiles are required, these values can be computed at the loading end, saving considerable computer effort.

8.5 Practical Application in Geomechanics

In this section, three geomechanical problems involving plastic behaviour are shown to illustrate the applicability of the initial stress technique in conjunction with the boundary element method. In the first illustration the unlined circular opening previously solved by Reyes (76) using finite elements is analysed with the boundary technique. The same example is used to simulate a tunnel in which the main axis is not a principal direction; in this case, the plastic solution is obtained assuming complete plane strain conditions. The third example consists of analysing the plastic solution for a lined tunnel with elliptical shape, previously solved by Valliappan (46). In all these applications only associative flow rules have been adopted, assuming also plane strain conditions with the exception of the second example where displacements at the tunnel axis direction are allowed.

(a)   Unlined Circular Opening

In this illustration an unlined circular opening is analysed using the initial stress process for boundary technique shown in the last section. The opening is deeply inserted in an intact rock domain which is assumed to be governed by the true Drucker-Prager criterion, i.e., the coefficients α and K are given by equation (8.3.15). The elastic modulus E and the Poisson's ratio ν are assumed to be 500 ksi and 0.2 respectively, while the yield function is characterised by the internal friction's angle $\phi = 30°$ and cohesion $c = 0.280$ ksi. The load applied to the tunnel surface is one due to the relief of the residual compressive state of stress. The residual vertical stress σ_v has been taken equal to the constant value of -1.0 ksi due to the depth of the tunnel, while two different values of the horizontal stress σ_h have been considered to account for the coefficient of earth pressure at-rest, K_o, taken equal to 0.25 and 0.4. The boundary discretization used to solve the problem is shown in fig. (8.5.1) in which the internal cells for the integration of the initial stress term are also presented. Plastic zones on complete unloading of the tunnel surface are presented in fig. (8.5.2), which shows very extensive yielding for K_o equal to 0.25 as was previously obtained in reference (76), using the finite element formulation. Normal and tangential stresses along the horizontal section are presented for both cases, $K_o = 0.25$ and 0.4, and the results are also in agreement with the previous finite element solution (fig. 8.5.3).

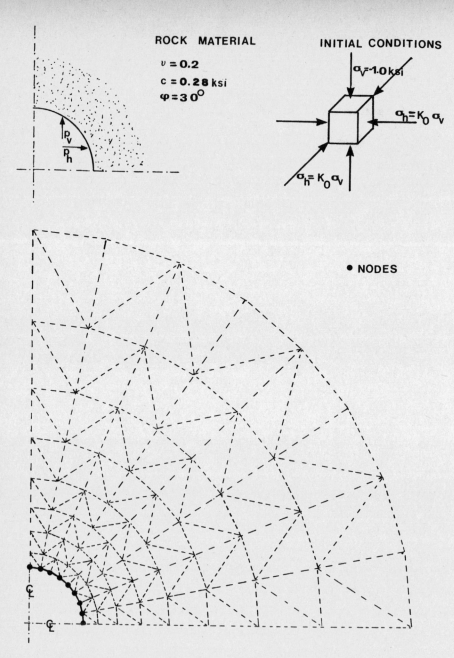

Figure 8.5.1 Circular Opening. Discretizations.

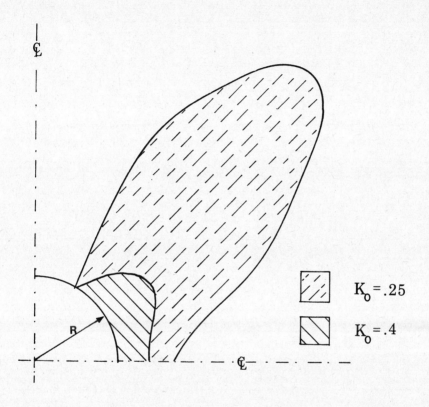

Figure 8.5.2 Plastic Zones.

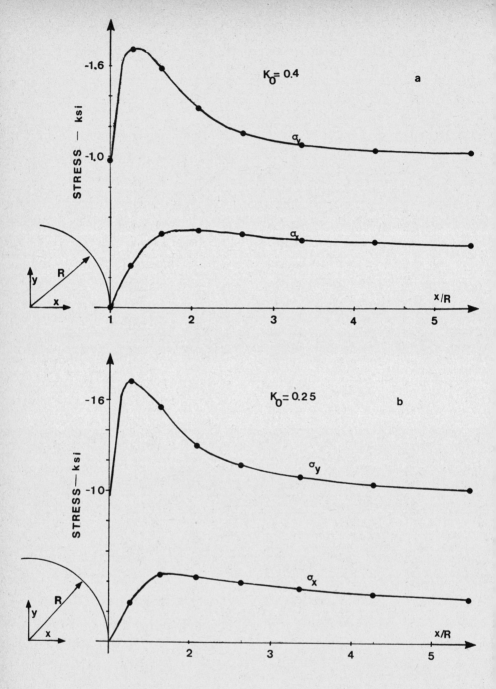

Figure 8.5.3 Stress Along the Horizontal Section.
(a) KO = 0.4 (b) KO = 0.25

For the case corresponding to K_o equal to 0.4, the stress distribution using the boundary element technique has already been presented in reference (39), where a similar result was obtained. Figure (8.5.4) shows elastic and plastic displacements of the tunnel surface for K_o equal to 0.25. These results also compare well with values given in reference (76).

(b)   Circular tunnel assuming complete plane strain conditions

The second example chosen to illustrate the use of the boundary element formulation to solve plastic problems is concerned with the application of the complete plane strain case presented in Chapter 4. The same circular tunnel already studied in the first example is assumed now to be excavated in a different situation; the opening axis is characterized by an angle equal to $15°$ related to any horizontal plane, as is shown in figure (8.5.5). The same residual stress field is assumed, and only the case in which the earth pressure coefficient at-rest, K_o, equal to 0.4 is taken into consideration. As the plane orthogonal to the tunnel axis is no longer related to the residual stress field, these stresses are properly computed to define initial condition and boundary tractions for the plane and anti-plane cases. All geometric parameters and rock properties have been assumed to be equal to those used in the first example. Also, boundary and internal discretization adopted in this case is the one presented in figure (8.5.1). The plastic zone obtained in this case (see fig. 8.5.6) shows the influence of the direction of the tunnel axis. The final displacements for plane and anti-plane case are presented in figure (8.5.7). In all directions the plastic flow leads to larger displacements in comparison with the initial elastic values. Figure (8.5.8) shows the stress distribution along the x-coordinate axis. The results are slightly different from

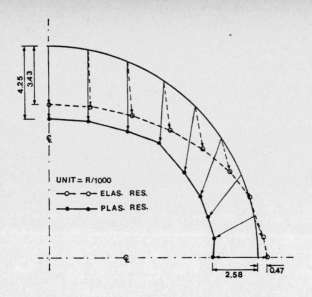

Figure 8.5.4 Elastic and Plastic Displacements on the Tunnel Surface.

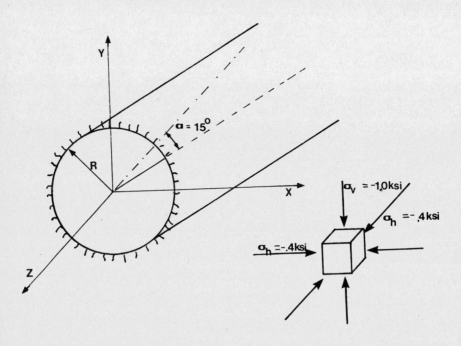

Figure 8.5.5 Tunnel with Axis not Coincident with a Principal Direction.

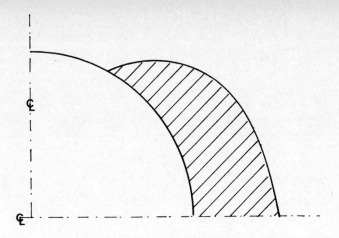

Figure 8.5.6 Plastic Zone. Complete Plane Strain Case.

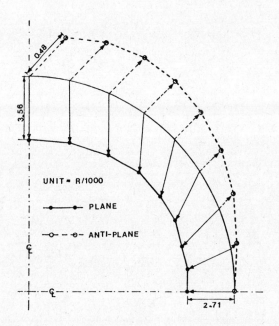

Figure 8.5.7 Boundary Displacements. Complete Plane Strain Case.

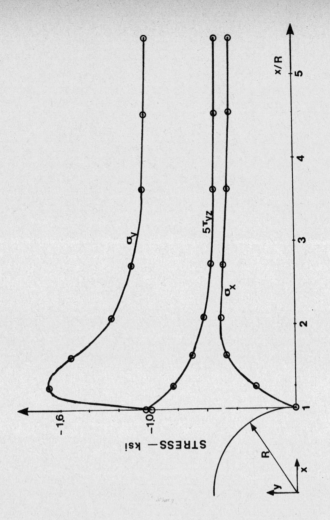

Figure 8.5.8 Stress Distributions along x-Axis.

the previous example due to both the orientation of the residual stresses and anti-plane consideration.

(d)   Lined Tunnel

In this application an elliptical lined tunnel taken from ref. (46) is analysed. The concrete lining support is assumed to behave elastically whereas the surrounding rock is considered to obey a yield function. Two plastic criteria (Mohr-Coulomb and the true Drucker-Prager) have been chosen, in this case without any hardening effects, i.e., elastic perfectly plastic conditions have been assumed for the rock. The tunnel is placed at a depth of 330 ft and the material is initially compressed due to its self-weight taken equal to 150 lb/cu.ft. The coefficient of earth pressure at-rest, K_o , which gives the horizontal compressive stress field, is assumed to be 0.2.

The material properties for concrete and rock are given below (table 8.5.1).

Parameter and properties	Rock	Concrete
Young Modulus	5×10^5 psi	3×10^6 psi
Poisson's ratio	0.2	0.15
Material behaviour	Plastic	Elastic
Cohesion	20000.0 lb/ft^2	-
Friction's angle	30°	-
Uniaxial slope	0.0	-

Table 8.5.1 Material Properties

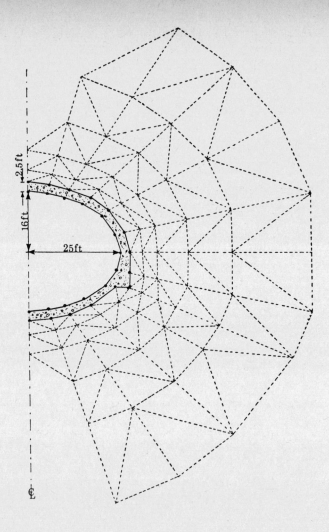

Figure 8.5.9 Lined Tunnels. Geometry and Discretizations.

The dimensions of the problem are given in fig. (8.5.9) where boundary element and internal cell configurations are also presented. The applied load on the interface between concrete and rock is due to the removal of the initial compressive state of stress around the opening existing before the excavation. No displacements are assumed to take place before the completion of the lining; this is not a realistic assumption, but is justified when the lining is built at a very short distance from the excavation front. Figure (8.5.10) shows the plastic zones developed during the loading process. The total displacements in the internal lining face are presented in fig. (8.5.11) for both Mohr-Coulomb and Drucker-Prager criteria. In spite of different yield zones developed for each criterion, the displacement profiles are very similar. In figure (8.5.12) the tensile zones for the Drucker-Prager yield criterion are also presented.

As the solution presented in reference (46) did not agree with the results obtained here, a finer mesh defined by 50 boundary elements and 185 internal cells was also employed, but it was found that all values remain practically unchanged, what seems to indicate some error in the results previously published. Similar but not identical example has been also solved in reference (18) where the same patterns for plastic zones, displacements and tensile regions were obtained using a viscoplastic algorithm.

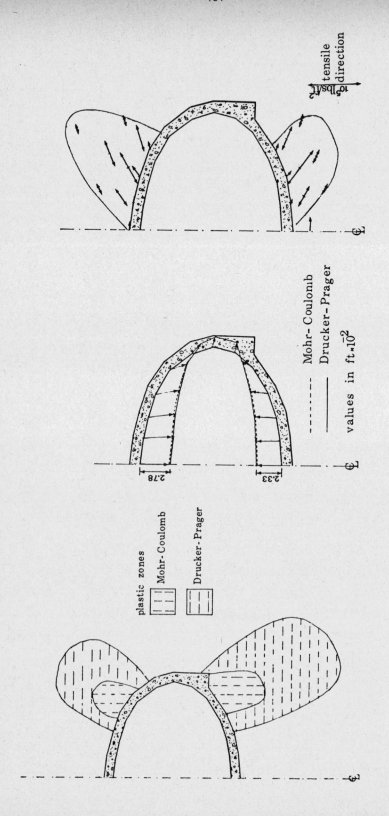

Figure 8.5.10 Lined Tunnel Plastic Zones.   Figure 8.5.11 Lining Displacements   Figure 8.5.12 Tensile Zones

CHAPTER 9

ELASTO/VISCOPLASTIC BOUNDARY ELEMENT APPROACH

9.1 Introduction

In this chapter a boundary element approach to deal with continuum problems involving viscous effects is presented. Basic concepts of time-dependent material behaviour, together with some rheological models used in geotechnical engineering, are introduced for one-dimensional cases. The Perzyna's viscoplastic model is chosen to model viscoplastic behaviour. Based on this viscoplastic representation, the theory is extended to general continuum problems. The overlay concept borrowed from finite elements formulation (48,49) is also adopted here in order to model more complex time-dependent and plastic responses. As in finite elements, the technique can be also extended to solve problems with pure plastic behaviour. Finally, the applications presented in the last section of this chapter show the accuracy of the approach when it is used to model time-dependent behaviour or to simulate plastic responses with or without anisotropic effects.

9.2 Time-Dependent Behaviour in One Dimension

For the analysis of stresses and displacements of any soil or rock structural system, the material properties must be considered as realistically as possible in order to provide safe and stable construction. The plasticity and no-tension theories previously presented are based on assumptions in which the time properties of the material are fully neglected; however, it is known that in many geomechanical problems the actual behaviour of the material is also governed by rheological properties. For instance, plastic solutions are acceptable when the onset of the plastic deformations occurs much faster than the loading time, i.e., the concept of instantaneous development of permanent and irreversible strain is valid. On the other hand, the viscous effects become important when permanent deformations take a comparatively long time to be developed.

Many kinds of rock and soil materials show distinct time-dependent behaviour. Generally, these material behaviours are governed by constitutive laws mathematically formulated relating stress and strain tensors or their derivatives with reference to time. As in the plasticity theory, the time-dependent concept is best introduced by reference to the one-dimensional case. Then, the creep deformation, which can be seen for many materials as the most important time-dependent behaviour, is generally described by the curves in figure (9.2.1). On applying constant stresses on a uniaxial specimen, instantaneous deformations characterized by the lines $A-B_i$, $i = 1,2,3$, take place, then creep deformations occur as a function of the applied stresses. The first part of the three curves (lines B_i-C_i, $i = 1,2,3$) is characterized by rates which decrease with increasing time; this time-dependent deformation, often called primary creep or viscoelasticity, is recoverable by removing the applied load. For stresses below a certain limit, it is the only kind of time-dependent response obtained and the strain tends asymptotically to a final value (see line B_1-C_1). If the stress exceeds the limit mentioned above, a second kind of time-dependent deformation will follow the primary creep phase. This behaviour, called secondary creep, shows a constant rate (see curve C_2-D_2) and the deformations are not recoverable when the loads are removed. Another creep phase, named terciary creep, can also occur. This kind of time-dependent deformation is illustrated by line D_3-E_3

and characterized by an increasing rate leading to fracture of the specimen.

Another time-dependent behaviour often assumed for soil and rock material is given within the viscoplasticity concept. In this case the material is assumed to behave elastically when the applied stress is below a certain limit and shows viscous response under action of higher stresses. The main characteristic of this behaviour is the irreversibility of all deformations.

It is usual to represent all these time-dependent behaviours with either the assumption of some idealized models or a combination of them. The two most common viscous responses assumed for soil and rock material, viscoelasticity and elasto/viscoplasticity, can be represented by the rheological models given in figures (9.2.2) and (9.2.3) respectively. For the first unit two spring and one dashpot are used. The spring placed in series gives the initial elastic response, while the dashpot and the other spring govern asymptotic displacements with reference to time and the total viscous deformation respectively. In the case of the viscoplastic representation, the spring gives the elastic response, and the stress level is governed by the slider which allows viscous deformation only for stress greater than a certain limit.

In this work, the Perzyna's elasto/viscoplastic model (16, 107, 108 and 109) is adopted. Therefore the stress conditions, which indicate the onset of viscoplastic deformations, are governed by a well known plastic yield function, i.e., irreversible viscous deformations take place only for,

$$F(\sigma, k) > 0 \qquad (9.2.1)$$

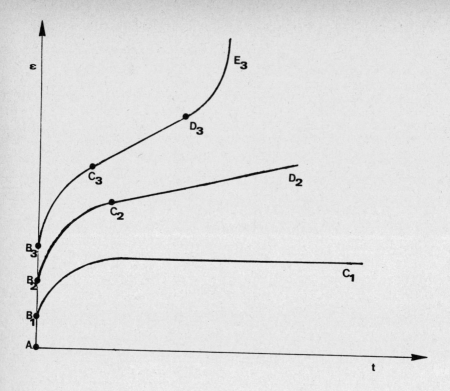

Figure 9.2.1 Uniaxial Creep Curve Under Constant Load.

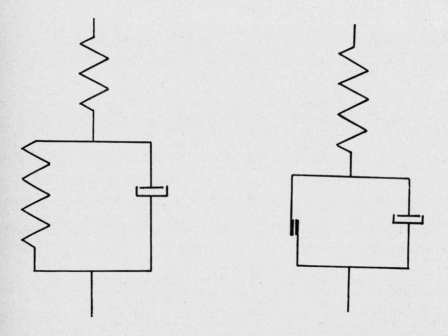

Figure 9.2.2 Viscoelastic Model   Figure 9.2.3 Elasto/Viscoplastic Model.

where the function F represents a yield criterion as was presented in section (8.2).

In the usual manner for nonlinear problems, the total strain can be separated into elastic and inelastic parts. In this case in particular, the total strain rate, $\dot{\varepsilon}$, can be expressed as,

$$\dot{\varepsilon} = \dot{\varepsilon}^e + \dot{\varepsilon}^{vp} \qquad (9.2.2)$$

in which $\dot{\varepsilon}^e$ and $\dot{\varepsilon}^{vp}$ stand for the elastic and viscoplastic strain rates, and

$$\dot{\varepsilon}^{vp} = \gamma <\Phi(F(\sigma, k))> \qquad (9.2.3)$$

where γ is used to denote the fluidity parameter of the material and $<\Phi>$ is a function relating the stress level and yield stress to be defined in the following section.

Equation (9.2.2) can also be presented in terms of strain increments by integrating the strain rates over an increment of time, i.e.,

$$\Delta\varepsilon = \Delta\varepsilon^e + \Delta\varepsilon^{vp} \qquad (9.2.4)$$

The elasto/viscoplastic behaviour can also be represented by the model illustrated in figure (9.2.3) in which viscous effects given by the dashpot are governed by both the viscous parameter and yield function. The model gives a convenient representation for the secondary creep when no stress limit is associated with the slider. In this case a Maxwell unit is formed and a constant strain rate is modelled. Other models can be obtained by combinations of two or more viscoplastic units. One such combination is obtained by the use of "overlay models" (49) which enables individual units to be placed in parallel in order to model more complex material behaviour.

The overlay concept already mentioned in section (8.2) is based on the assumption that the body is formed by several layers or overlays, as shown in figure (9.2.4). Different material properties, together with a weighted parameter, namely overlay thickness, are assigned to each overlay. By enforcing the same strain pattern in each layer, different stress responses are obtained, which are employed to compute the final values as follows,

$$\sigma = {}^{(m)}\sigma\, h_m \qquad (m = 1,\ldots,M) \qquad (9.2.5)$$

where ${}^{(m)}\sigma$ stands for the stress level at each overlay, M is the total number of overlays adopted, h_m represents the overlay thickness, and

$$\sum_{m=1}^{M} h_m = 1 \qquad (9.2.6)$$

Using this association of models, many material behaviours can be readily represented by an appropriate choice of the material properties for each layer or model.

The viscoelastic rheological model given in figure (9.2.2) can be simulated by two overlays (fig. 9.2.5) represented by degenerated elasto/viscoplastic units. In the first unit an infinite yield stress is assigned and the corresponding overlay models only elastic response. For the second overlay, the yield stress is taken equal to zero leading to the Maxwell model, which combined with the first unit, can simulate viscoelastic responses. Another useful combination is given by two overlays as shown in figure (9.2.6), where one unit is degenerated

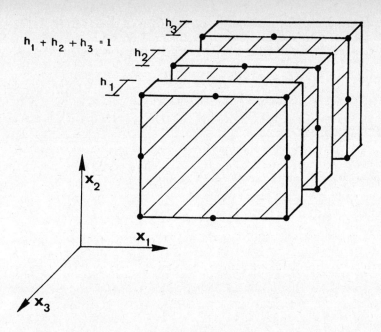

Figure 9.2.4 Overlay Representation.

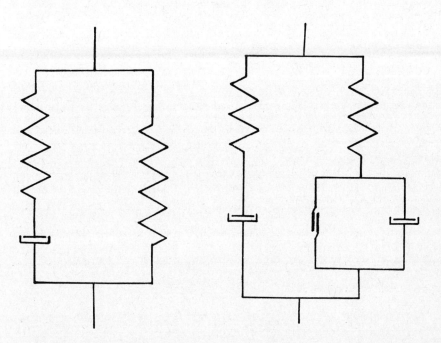

Figure 9.2.5 Viscoelastic Model.

Figure 9.2.6 Elasto/Viscoplastic Model.

to the Maxwell model. This model simulates both viscoelastic and viscoplastic behaviour. For stresses below the yield limit the model behaves as a viscoelastic unit, while viscoplastic are modelled when the slider yields. This representation appears to be useful in many geomechanical problems in which viscous responses due to the applied loads might occur everywhere, specially in the vicinity of the load, where the strain rates are greatly increased.

9.3 Elasto/Viscoplastic Constitutive Relations for Continuum Problems

As shown for the uniaxial case, the onset of viscoplastic deformation is also governed by a yield criterion for continuum problems i.e.,

$$F(\sigma_{ij}, k) = 0 \qquad (9.3.1)$$

which has already been described in the previous chapter.

As before, for practical purposes the yield criterion can be represented by,

$$f(\sigma_{ij}) - Y(k) = 0 \qquad (9.3.2)$$

in which $Y(k)$ is the yield stress, k is a history dependent hardening (or softening) parameter, and $f(\sigma_{ij})$ can be interpreted as an equivalent uniaxial value σ_e.

In order to extend equation (9.2.3) to continuum problems, the concept of plastic potential from the plasticity theory can be adopted and then the viscoplastic strain rate is written as follows,

$$\frac{d\varepsilon_{ij}^{vp}}{dt} = \gamma <\Phi(F/F_o)> \frac{\partial}{\partial \sigma_{ij}} g$$

or

$$\dot{\varepsilon}_{ij}^{vp} = \gamma <\Phi(F/F_o)> a_{ij}^{(G)} \qquad (9.3.3)$$

where g is a scalar plastic potential function also given in terms of the stress tensor (defined in section 8.3) and F_o denotes any convenient reference value of F, usually Y(K), for the dimensionless representation of Φ.

As no viscoplastic deformation occurs below the yield limit, the $< >$ notation in equation (9.3.3) implies,

$$<\Phi(F/F_o)> = \Phi(F/F_o) \quad \text{if} \quad F > 0$$

$$<\Phi(F/F_o)> = 0 \qquad \text{if} \quad F \leq 0 \qquad (9.3.4)$$

From the results of experimental tests, the function Φ may assume different forms, as suggested in reference (16), in which the following representations have been proposed.

$$\Phi(F/F_o) = F/F_o$$

$$\Phi(F/F_o) = (F/F_o)^n \qquad (9.3.5)$$

$$\Phi(F/F_o) = \exp(F/F_o) - 1$$

They correspond to linear, power and exponential types respectively.

After multiplying equation (9.3.3) by σ_{ij}, the same procedure used to derive (8.3.28) can be adopted to achieve (40),

$$\dot{\varepsilon}_{ij} \dot{\varepsilon}^{ovp} = \gamma <\Phi(F/F_o)> g(\sigma_{ij}) \qquad (9.3.6)$$

Recalling the concept of work hardening given by equations (8.2.6) and (8.3.24) for the uniaxial and continuum cases respectively, and applying them in their rate forms, the following expression is derived,

$$\dot{\varepsilon}_e^{vp} = \gamma \langle \phi(F/F_o) \rangle \qquad (9.3.7)$$

which stands for the equivalent viscoplastic strain rate.

The general form of equation (9.2.2) for a continuum problem can also be used to represent the total strain rate vector of a particular overlay in function of the elastic and viscoplastic strain rate components as follows,

$$^{(m)}\dot{\varepsilon}_{ij} = {}^{(m)}\dot{\varepsilon}_{ij}^e + {}^{(m)}\dot{\varepsilon}_{ij}^{vp} \qquad (9.3.8)$$

where the total strain rate vector must be the same in all overlays.

Multiplying equation (9.3.8) by the elastic compliances gives

$$^{(m)}\dot{\sigma}_{ij} = {}^{(m)}\dot{\sigma}_{ij}^e - {}^{(m)}\dot{\sigma}_{ij}^{vp} \qquad (9.3.9)$$

For practical purposes the incremental form of (9.3.9) obtained by integrating it over an increment of time Δt is usually adopted, as follows,

$$^{(m)}\Delta\sigma_{ij} = {}^{(m)}\Delta\sigma_{ij}^e - {}^{(m)}\Delta\sigma_{ij}^{vp} \qquad (9.3.10)$$

which can also be written as a function of the total values,

$$\Delta\sigma_{ij} = \Delta\sigma_{ij}^e - \Delta\sigma_{ij}^{vp} \qquad (9.3.11)$$

As the same strain pattern is enforced in all overlays the viscoplastic stress increment tensor, $\Delta\sigma_{ij}^{vp}$, is also computed following equation (9.2.5), i.e.,

$$\Delta\sigma_{ij}^{vp} = {}^{(m)}\Delta\sigma_{ij}^{vp} h_m \qquad (m=1,2,3,\ldots,M) \qquad (9.3.12)$$

in which M is the overlay number.

9.4 Outline of the Solution Technique

The procedure adopted for the viscoplastic solution is similar to those used to model no-tension and plastic responses presented in previous chapters, and consists of the following steps,

(i) Starting from known values of $^{(m)}\underset{\sim}{\sigma}^n$ and $^{(m)}F^n$, which represent the total stress vector and the yield function F for the overlay "m" at time t^n, the strain rate vector, $\underset{\sim}{\dot{\varepsilon}}^{vp}$, is computed by equation (9.3.3); then the corresponding stress rate vector, $^{(m)}\underset{\sim}{\dot{\sigma}}^{vp}$ is also evaluated.

(ii) Choosing a time step Δt and adopting a simpler Euler time integration scheme, the viscoplastic stress increment vector of one overlay is calculated as follows,

$$^{(m)}\Delta\underset{\sim}{\sigma}^{vp} = \int_t^{t+\Delta t} {}^{(m)}\underset{\sim}{\dot{\varepsilon}}^{vp} \, dt = {}^{(m)}\underset{\sim}{\dot{\varepsilon}}^{vp} \Delta t \qquad (9.4.1)$$

and the weighted value computed by,

$$\Delta\underset{\sim}{\sigma}^{vp} = {}^{(m)}\Delta\underset{\sim}{\sigma}^{vp} h_m \qquad (m = 1,2,\ldots,M) \qquad (9.4.2)$$

(iii) Applying the weighted viscoplastic stress increments, $\Delta\underset{\sim}{\sigma}^{vp}$, as an initial stress field, gives the total elastic stress increment vector,

$$\Delta\underset{\sim}{\sigma}^e = S\Delta\underset{\sim}{\sigma}^{vp} \qquad (9.4.3)$$

or the corresponding elastic value for each overlay,

$$^{(m)}\Delta\underset{\sim}{\sigma}^e = {}^{(m)}S\Delta\underset{\sim}{\sigma}^{vp} \qquad (9.4.4)$$

(iv) The total stress vector for one overlay "m" can now be obtained at time $t^{n+1} = t^n + \Delta t$ by accumulating the elastic and viscoplastic stress increment components, i.e.,

$$^{(m)}\sigma^{n+1} = {}^{(m)}\sigma^n + {}^{(m)}\Delta\underline{\sigma}^e - {}^{(m)}\Delta\underline{\sigma}^{vp} \qquad (9.4.5)$$

then the total stress vector for the whole body is computed once more using the weighted parameter,

$$\underline{\sigma}^{n+1} = {}^{(m)}\underline{\sigma}^{n+1} h_m \qquad (m = 1,2,\ldots,M) \qquad (9.4.6)$$

(v) Before computing the **next** time step, the new value of the equivalent plastic strain vector can be determined as follows,

$$\underline{\varepsilon}_e^{vp(n+1)} = \underline{\varepsilon}_e^{vp(n)} + \Delta\underline{\varepsilon}_e^{vp} \qquad (9.4.7)$$

in which $\Delta\underline{\varepsilon}_e^{vp}$ is computed employing the equivalent viscoplastic strain rate given in equation (9.3.7).

It is important to notice that the steps shown above only model the time behaviour starting from any state of stress of the body under consideration. In order to consider the application of any load, whether at the beginning or during the process, its elastic effects must be instantaneously added to both elastic and total stress vectors.

9.5 Time Interval Selection and Convergence

The success of the incremental technique given in the previous section for solving elasto/viscoplastic problems is directly dependent upon the appropriate time step selection. Then, in order to regulate the time step length, two procedures borrowed from finite element formulation are adopted in this work.

The first corresponds to the application of a limit time step formulated for creep problems (110), and adopted several times in geomechanics. The time step length is chosen in order to achieve accuracy in the results and is calculated as a function of the relation between the equivalent viscoplastic strain rate and the accumulated equivalent strain, ε_e, as follows,

$$\Delta t = \eta_o \frac{\varepsilon_e}{\dot{\varepsilon}_e^{vp}} \qquad (9.5.1)$$

where η_o is a constant to be chosen for each particular case; practical experience indicates the range $0.1 \leq \eta_o \leq 0.15$ for usual viscoplastic analysis, but near collapse or for stress concentration problems, lower values of the order of 0.05 or even smaller values have to be adopted.

This criterion tends to give small time step lengths at the beginning of the process in order to guarantee the accuracy of the results. However, it gives large steps when the steady state condition is approached. Thus a second experimental condition is usually adopted to limit the growth of the step length between two successive time steps as follows,

$$\Delta t = \eta_1 \Delta t \qquad (9.5.2)$$

where η_1 is usually chosen within the range $1.0 \leq \eta_1 \leq 1.5$.

In spite of the restriction enforced by equation (9.5.2), absolute guarantees regarding the propagation of errors are not achieved for the explicit time integration scheme proposed in the last section. In order to avoid instability problems, the time step length proposed by Cormeau (111) has been adopted here. Following Cormeau, the time marching procedure has also to obey a step limit which is evaluated based on material properties, the adopted yield function and the viscoplastic constitutive law.

For von Mises, Tresca and Mohr-Coulomb yield surfaces, the stable step limit is given by,

$$\Delta t_{max} = \frac{4(1+\nu)}{3\gamma E} F_o$$

$$\Delta t_{max} = \frac{1+\nu}{\gamma E} F_o \qquad (9.5.3)$$

$$\Delta t_{max} = \frac{4(1+\nu)(1-2\nu)F_o}{\gamma E(1-2\nu+\sin^2 \phi)}$$

respectively, when a linear type of viscoplastic function has been used (see equation 9.3.5).

For the Drucker-Prager surfaces the step limit is established according to two expressions, as follows,

$$\Delta t_{max} \leq \frac{4(1+\nu)}{E} \frac{\sqrt{J_2}}{\Phi} \qquad (9.5.4)$$

and

$$\Delta t_{max} \leq \frac{4(1-\nu)}{\gamma E} \frac{F_o}{\left[\beta^2 + \alpha^2 \left(\frac{1+\nu}{1-2\nu}\right)\right]} \qquad (9.5.5)$$

in which β and α are given for each particular Drucker-Prager surface. For the true Drucker-Prager, they are expressed by,

$$\alpha = \frac{\sin\phi}{(3+\sin^2 \phi)} \quad ; \quad \beta = \sqrt{3} \qquad (9.5.6)$$

An alternative to Cormeau's stability criterion is given in reference (40), where the stability condition for the Euler procedure has been obtained by analysing only equivalent values of the stress tensor instead of using their components.

As has been known for some time, the viscoplasticity algorithm can be used to model pure plastic behaviour by applying the load in increments and allowing the solution to progress until stationary conditions are attained (112). After applying an increment of load the steady state solution which corresponds to the condition $F \leq 0$ is approached asymptotically with the viscoplastic strain rate diminishing as the yield surface is approached. As for plastic solution, the condition $F \leq 0$ must be attained during the whole loading process; the accuracy of such an approach is directly dependent on the increment lengths adopted.

The steady state situation mentioned above is only obtained for a finite time if a tolerance parameter to control the accuracy of the results is specified. In the applications discussed in the following section and in Chapter 10, the steady state solution is considered achieved when the relation F/F_o is smaller than a tolerance taken to be 1/1000 or 1/500.

9.6 Elasto/Viscoplastic Applications

In this section some numerical analyses are shown in order to demonstrate the applicability of the elasto/viscoplastic approach presented. The circular tunnel analysed in the last chapter with the plastic algorithm are studied here using the elasto/viscoplastic technique. Two other examples unrelated to geomechanics are also presented in order to illustrate the overlay concept in the boundary element formulation.

(a)   Circular Tunnel.

The circular tunnel analysed in the previous chapter is discussed again here, assuming that the rock is an elasto/viscoplastic material. Only the case of $K_o = 0.40$ for both plane strain and complete plane strain conditions has been analysed. All material properties and the yield surface adopted were taken from the plastic case. The visco responses are considered to take a long time in order to admit the assumption that viscoplastic strains only take place after the completion of the opening. As only the final steady state situation is sought, the viscosity parameter and the time step were chosen according to the limit given in equations (9.5.4 and 9.5.5).

Due to the nature of the problem, all the results show small changes in relation to the previous plastic solutions. For this particular example, one can conclude that the development of plastic deformations is not strongly influenced by the load path. Despite all of the loads being applied in one increment, the plastic regions obtained in this analysis are practically those already presented in figures (8.5.2) and (8.5.6). The final displacements computed for both plane strain and complete plane strain analyses show an increase of around 0.5%, which also illustrates the small effects introduced by the incremental procedure in this particular case. The stresses along a horizontal axis, already shown in figures (8.5.3) and (8.5.8), were almost unaffected as well. Maximum changes occur in the stress peak values, which were decreased by about 2%.

(b)   Perforated Strip

The inclusion of such an example in this work is to illustrate the use of the overlay technique in conjunction with boundary formulations.

The viscoplastic algorithm is applied in order to model a pure plastic solution for a perforated plate under tensile loading. As has already been mentioned, the plastic solution is obtained by applying the load in increments in which steady state conditions have to be achieved.

The example was taken from reference (48), in which a finite element solution is presented. Geometric dimensions, boundary and internal discretizations used to solve the problem are shown in figure (9.6.1) together with the mesh adopted for the finite element analysis. The elastic material constants adopted for the analysis are,

$$E = 10000 \text{ kgf/mm}^2$$

$$\nu = 0.3$$

and the uniaxial stress-strain curve chosen to govern the plastic behaviour after yielding is given in figure (9.6.2). The elastic limit σ_y is taken equal to 10 kgf/mm^2, while the maximum yield stress value is σ_u 20 kgf/mm^2. For the analysis, the stress-strain curve is approximated by straight segments corresponding to 5 overlays. Each overlay is characterized by its material properties and its thickness. The thickness or weighted parameter can be defined according to the linear approximation chosen to represent the stress-strain curve (fig. 9.6.2), i.e.,

$$t_i = \frac{\tan\alpha_i - \tan\alpha_{i+1}}{E} \qquad (9.6.1)$$

In this case all overlays are assumed to have the same elastic constants (E and ν) already defined for the body. Uniaxial elasto perfectly plastic behaviour is also associated with the overlays, and their yield stresses are obtained according to the actual stress-strain curve (fig. 9.6.2), i.e.,

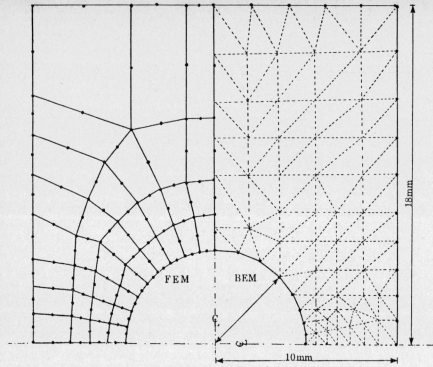

Figure 9.6.1 Perforated Plate. Geometry and Discretizations.

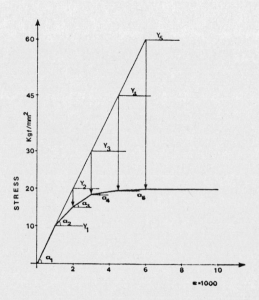

Figure 9.6.2 Uniaxial Stress-Strain Curve and the Overlay Representation.

$$Y_i = E\varepsilon_i \qquad (9.6.2)$$

Therefore, using equations (9.6.1) and (9.6.2) and adopting the linear approximation given in figure (9.6.2), the following characteristics are defined for the overlays,

overlay	1	2	3	4	5
thickness	0.5	0.3	0.0667	0.0667	0.0666
yield stress (kgf/mm^2)	10	20	30	45	60

The von Mises yield criterion was chosen to model the plastic behaviour of each overlay, and also the time marching procedure adopted follows equation (9.5.3). The load was incrementally applied in the form of constant displacements applied over the plate's end. The development of the plastic zones according to the displacement applied is illustrated in figure (9.6.3), and it is seen that practically identical patterns are shown for finite (48) and boundary element solutions. Figure (9.6.4) shows the total reaction on the plate's end as a function of the displacements applied, and once more boundary and finite element techniques compare well.

Assuming the same material characteristics given in figure (9.6.2), the case of cyclic uniaxial loading has also been analysed. Using a square plate with four boundary elements, the Bauschinger effect is obtained. After an initial tensile loading it is clear that the onset of yield in compression is dependent on the previous tensile stresses, as illustrated in figure (9.6.5).

(c)   Polystyrene Crazing Problem

This is another example solved in order to emphasize the use of overlay models together with boundary element formulation. The problem

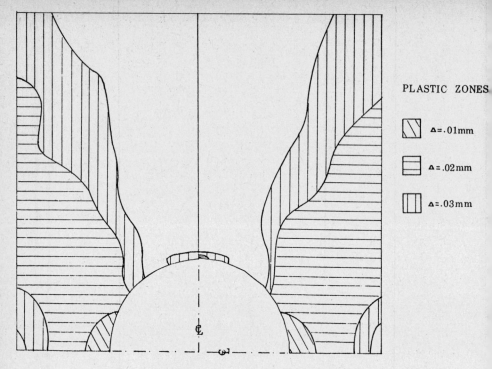

Figure 9.6.3 Plastic Zones as a Function of the Prescribed Displacement Δ.

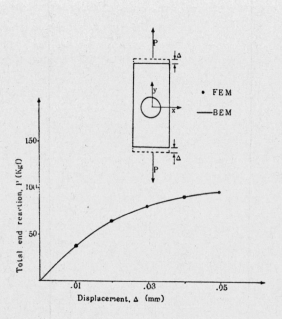

Figure 9.6.4 Reaction-Displacement Curve.

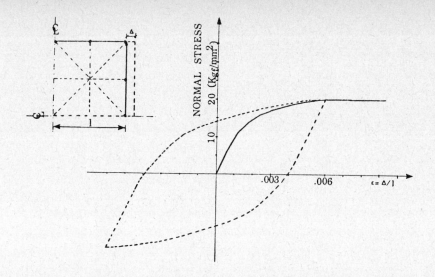

Figure 9.6.5 Uniaxial Loading With Bauschinger Effects Modelled.

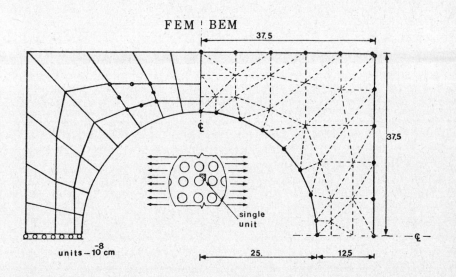

Figure 9.6.6 Polystyrene Crazing Problem. Geometry and Discretizations.

was also run by Owen, Prakash and Zienkiewicz (48), using finite element methods, and consists of analysing the voids formed in polystyrene under tensile loading. The geometry of the body is defined in figure (9.6.6), where discretizations for finite and boundary element analyses are presented. As shown in this figure, the discretized domain was taken to model an infinitely extended array of circular holes in a two-dimensional medium. The uniaxial stress-strain curve for the material shows both strain-hardening and strain-softening behaviours, as can be seen in figure (9.6.7). For both finite and boundary element analyses, this curve was represented piecewise linearly, which is simulated by the overlay approach. The strain-softening characteristic shown in the uniaxial stress-strain curve is modelled by adopting negative thicknesses. Five overlays associated with elasto perfectly plastic material were used. The elastic constants employed were,

$$E = 4.3 \times 10^7 \text{ gf/cm}^2$$

$$\nu = 0.33$$

while the overlays are defined by the following thicknesses and yield stresses,

overlay	1	2	3	4	5
thickness (t_i)	0.8432	0.2944	-0.1373	-0.0114	0.0111
yield stress (Y_i) (10^6 gf/cm^2)	0.85	1.4515	4.30	17.20	35.26

Figure (9.6.8) shows the plastic zones developed for finite and boundary element analyses and good agreement is verified. The load-deformation characteristics obtained using boundary formulation are given in figure (9.6.9) and they also compare well with the previous finite element results (48).

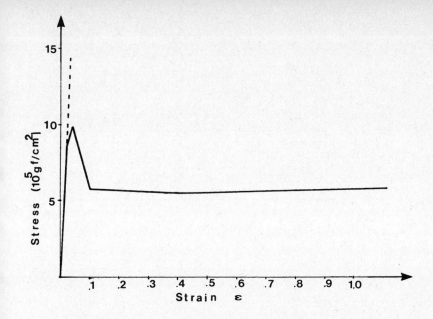

Figure 9.6.7 Uniaxial Stress-Strain Curve.

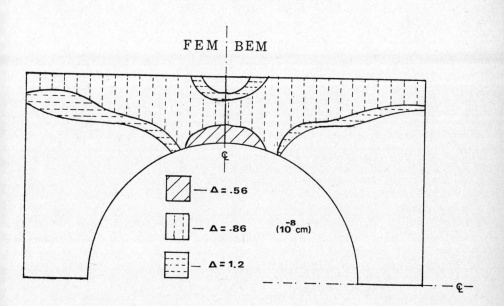

Figure 9.6.8 Plastic Zones Developed During the Loading Process.

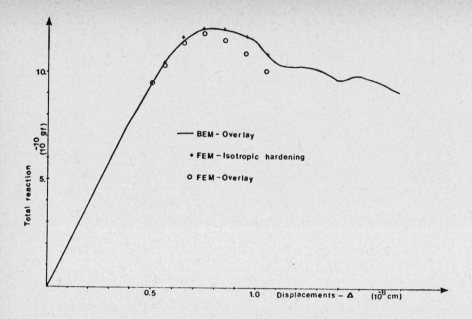

Figure 9.6.9 Load-Displacement Curve for Boundary and Finite Element Approaches.

CHAPTER 10

APPLICATIONS OF THE NONLINEAR BOUNDARY ELEMENTS FORMULATION

10.1 Introduction

In this chapter some geomechanical nonlinear problems are analysed using the boundary element technique presented in the previous chapters. For all examples the viscoplastic algorithm shown in chapter 9 is employed, even when only plastic solutions are analysed.

The first example is concerned with the determination of ultimate capacity of a single strip footing bearing on a plane surface of a semi-infinite mass of soil that is assumed to be elastic perfectly plastic material. Initially the problem is solved neglecting the weight of soil and then the analysis is extended to the case in which the unit weight of the material is considered.

Embankment and excavation results are presented in the second application. In both cases the slope stability is analysed with viscoplastic solutions. The loads are applied in one increment and the final displacements are computed when the stationary conditions are achieved. Then, the limit load is given by the last value of the load for which a steady situation is reached.

In the last example a lined circular tunnel excavated in a uniform stressed solid rock is investigated, taking into account the real time-dependent behaviour of the material. Several loading conditions are considered trying to model important stages of the tunnel in both the excavation time and the operation period.

10.2 Strip Footing Problem

In this example the ultimate capacity of a 10ft wide flexible strip footing resting on the surface of elastic perfectly plastic soil is analysed. The material was assumed to have an elastic modulus of 4.82×10^6 lb/ft^2, a Poisson's ratio equal to 0.3, and to be isotropic and homogeneous. The plastic deformations were assumed to be governed by the Mohr-Coulomb criterion with associative flow rule for which the following material parameters were used,

$$c = 1440 \text{ lb/ft}^2$$
$$\phi = 20°$$

The final plastic solution was obtained by using the viscoplastic algorithm. The load was applied in small increments and the convergence was assumed when the stationary conditions were achieved. The time step and the viscosity parameter have been chosen in order to obey the stability conditions given in section (9.5).

(a) Weightless soil case

We shall start the analysis of the strip footing by neglecting the soil weight. The boundary and internal discretizations employed to run the problem are shown in figure (10.2.1) together with the boundary conditions prescribed in this case. In common with finite element analysis for this example (see reference 17) a closed domain has been chosen. This kind of restriction for which the displacements are enforced to be zero at points not very distant from the applied load is a normal procedure in finite element techniques and has proved to be acceptable for the bearing capacity determinations (113).

The collapse load predicted by Prandtl (51) was simulated with 1% of error. The load increment for which the convergence was verified gives the total amount of 21.600 lb/ft^2, while the theoretical solution is 21.370 lb/ft^2.

The problem has also been solved with the associated true Drucker-Prager yield criterion and the same bearing capacity was obtained. For instance, the definition of the parameters α and K for this criterion (eq. 8.3.15) was established in order to reproduce the Mohr-Coulomb hypothesis for plane strain conditions (12).

The load displacement curves for both criteria are shown in figure (10.2.2) where one can see that the displacements for the Drucker-Prager criterion are larger than the corresponding values computed using the Mohr-Coulomb surface.

The growth of the plastic zone defined by the Mohr-Coulomb criterion is shown in figure (10.2.3). It began at points underneath the load and only spread all over the domain when almost all increments of load had been applied.

The determination of the limit load has also been carried out by analysing the elasto/viscoplastic response. The load is applied in one increment and the final solution is obtained when the stationary conditions are achieved. By solving the problem several times it was found that a load equal to 21600 lb/ft^2 was the largest value for which the steady solution was obtained. For any load greater than this value, a continuous growth of the displacements with respect to time was observed. In spite of being a non incremental procedure, the displacements computed with this procedure are similar to those obtained following the incremental scheme. Significant differences have been verified only near the limit load, as shown in figure (10.2.2).

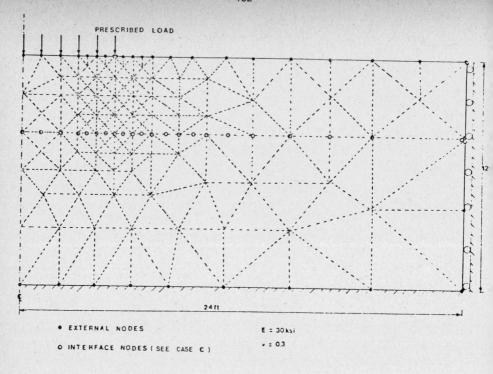

Figure 10.2.1 Strip Footing. Problem Definition and Discretizations.

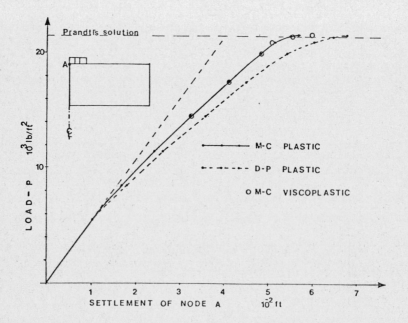

Figure 10.2.2 Load Displacement Curve.

The results presented above compare well with the finite element solution for the same problem presented in reference (17). This example has also been solved by Telles and Brebbia (114) using an elastoplastic boundary element formulation where the Melan (32) fundamental solution was employed. The solution they have obtained is also in agreement with the presented results.

(b)   Bearing capacity considering the weight of the soil

This example represents a more realistic situation if compared with the previous case. For instance, it can represent any foundation resting on the ground surface. The self-weight of the soil $\bar{\gamma} = 120$ lb/ft^3 is taken into consideration by assuming a pre-existing stress field in the material. The vertical stresses are assumed to be equal to the soil weight while the horizontal values are computed considering the earth pressure coefficient at-rest, K_o, which is given in this case by $\nu/(1-\nu)$.

The same type of analysis carried out for the weightless case has been repeated here. A limit load equal to 24200 lb/ft^2 has been achieved and is much smaller than the collapse load computed using the Prandtl's mechanism, which is 25080 lb/ft^2. However, the collapse load assuming Prandtl's mechanism is rigorously an upper bound. In this case, the ultimate capacity achieved by adopting the Hill mechanism (50) and assuming valid the Terzaghi superposition (52) is 23450 lb/ft^2, only 3% smaller than the numerical result and can be considered a more realistic solution. Spencer (115) has also achieved a collapse load for this problem by using a perturbation technique, and his solution, 23800 lb/ft^2, is less than 2% smaller than the numerical limit obtained. Assuming the Drucker-Prager criterion as for case "a", the limit load obtained was virtually the same.

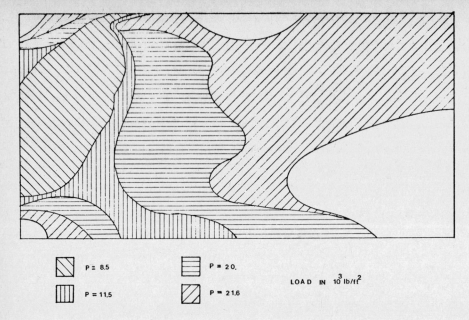

Figure 10.2.3 Plastic Zones.

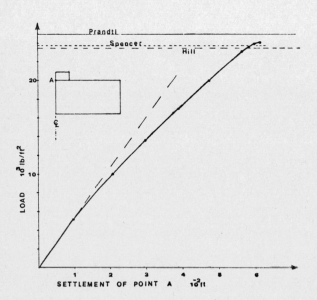

Figure 10.2.4 Load Displacement Curve Considering the Weight of Soil.

The load displacement curve for Mohr-Coulomb yield criterion is shown in figure (10.2.4) where the theoretical limit loads mentioned above are also presented. The plastic zone developed during the loading is presented in figure (10.2.5). By comparison with the first case, the final yield zone is smaller, which can be justified by the compressive state of stress assumed to exist before the loading.

(c)   Layered Soil Case

The reason for solving a half-plane problem employing Kelvin's solution is to show the ability of the technique to deal with practical cases concerned with geomechanical analysis. The Melan solution employed in reference (114), in order to study case "a" has proved to be efficient and the limit load was simulated with good accuracy using coarser mesh by comparison to the one presented in figure (10.2.1). The main reason for this is due to the elimination of the surface displacement interpolation. However, this kind of solution is of limited use for practical purposes. In most problems in geomechanics the assumption of horizontal and free ground surface cannot be made, leading the analysis to discretize the whole surface of the body; furthermore, the common case of layered soils requires discretization of internal interfaces in order to take into account the properties of different materials involved. Therefore, the applicability of the boundary element formulation to solve nonlinear problems in geomechanics is clearly dependent on the reliability of both Kelvin's fundamental solution and subregion technique. In cases "a" and "b" already presented, the limit loads for a strip footing neglecting or not the weight of soil, have been accurately represented. Now case "a" will be analysed again, this time considering an interface 4 ft underneath the ground surface. It must be noticed that this discretized surface will enforce displacement and traction interpolations in the zone characterized by large plastic deformations.

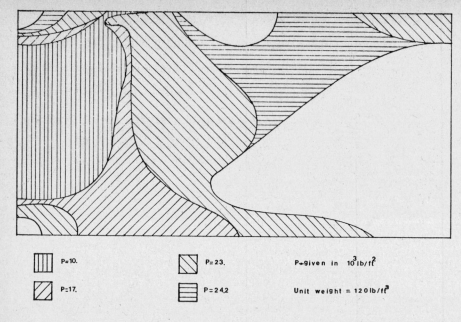

Figure 10.2.5 Plastic Zones.

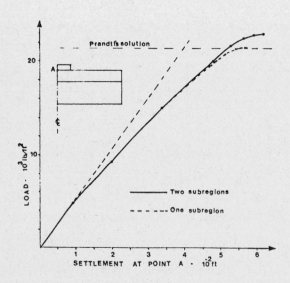

Figure 10.2.6 Load Displacement Curve. Subregion Case.

The interface nodes used to run this example are shown in figure (10.2.1). New cells have also been defined in order to avoid nodes on any triangle's straight edge.

The growth of the plastic zone observed for this case has been very similar to the previous results presented in figure (10.2.3). The last steady solution has been achieved for a load equal to 22.800 lb/ft^2, which is 5% larger than the previous value obtained. Figure (10.2.6) shows the load displacement curve obtained from this case in comparison with analogous result computed in "a".

(d)   Infinite Domain Case

Proper modelling of infinite domain is one of the boundary element technique's characteristics. In this example the strip footing already solved using a finite domain is now analysed taking into account the actual half-space. The ground surface is the only boundary to be discretized and it has been extended to 300 ft from the load region in order to simulate the free surface. The same internal discretization has been adopted, but now it is increased in order to take into account the expected large plastic zones, and only the case with residual compressive state of stress (as in "b") has been under consideration.

As expected, the load displacement curve for the elastic analysis shows that the displacements of the ground surface occur in the direction of the outward normal. This is consistent with the boundary element formulation in which only relative displacements have physical meaning. However, the plastic displacements computed by the analysis do not show the same pattern; the plastic settlements occur in the load direction, as illustrated in figure (10.2.7). Figure (10.2.8) shows

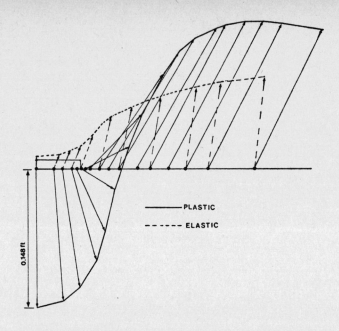

Figure 10.2.7 Displacements at Surface Nodes.

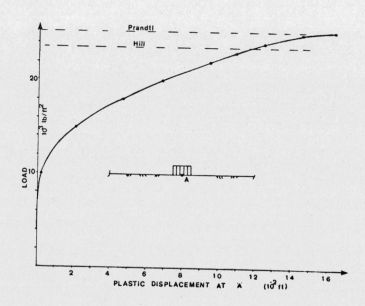

Figure 10.2.8 Load Displacement Curve. Infinite Domain Case.

the load settlement curve for a ground surface point on the centre line. As can be seen, very large plastic displacements are computed in comparison with the values obtained for the limited domain (case "b"). Although a coarser mesh for the extended part of the domain has been adopted, a limit load is also achieved, but it is larger than the previous value obtained considering closed domain.

10.3 Slope Stability Analysis

The example presented in this section concerns the application of the boundary element technique to the analysis of the earth slope stability. Most of the methods used for this purpose are based on the assumption of the existence of a failure surface, and the stability is verified according to the limit equilibrium. Once such technique is the conventional limit equilibrium approach, known as the "Swedish circle method", developed by Fellenius (118), in which a circle is chosen as the failure surface. Alternatively, the assumption of a logspiral failure surface for the analysis of slope stability is often employed (see ref. 53). Although the methods mentioned above have been adopted for many practical applications, realistic analysis considering the actual behaviour of the material can now be provided by numerical techniques.

The example chosen for this application was taken from reference (17), where it has been solved with a finite element code. Only two cases of the original example will be analysed here: an excavation in a uniform soil and an embankment construction. Viscoplastic algorithm has been applied for both analyses. The load is applied in only one increment and the failure was induced by studying the gradual reduction of the cohesion, while the angle of internal friction was maintained at a constant value.

(a)   Excavation

For this example a 10 m deep excavation, as shown in figure (10.3.1), is analysed. The domain is assumed to be infinite and the ground surface was discretized into linear elements up to a distance large enough to simulate the infinite boundary, as shown in figure (10.3.1). The domain discretization was also chosen to allow integration of the inelastic strain over a convenient area.

The residual compressive state of stress assumed to exist before the excavation is computed according to the unit weight of the material, $\bar{\gamma} = 2.0$ tf/m^3, and the distance of each point to the actual ground surface. The earth pressure coefficient at-rest, K_o, has been adopted as in reference (17), i.e., $K_o = 0.65$, in order to compute the horizontal stresses. Using these residual stress values, the resulting loads can be computed and applied on the excavated surface.

The material parameters used for this application were chosen as follows,

$$E = 2.0 \times 10^4 \text{ tf/m}^2$$

$$\nu = 0.25$$

$$\phi = 20°$$

Yield Criterion = Associative Mohr-Coulomb

As mentioned before, in this analysis the friction angle is maintained constant, and the stability of the slope is studied for the variation of the cohesion value. Therefore, the results are presented in the form of cohesion displacement curves. Figure (10.3.2) shows the vertical displacements of a particular point. An equivalent curve for

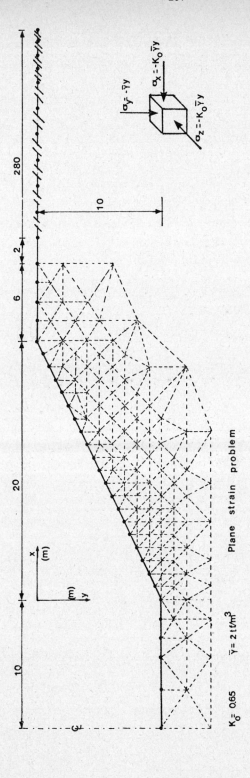

Figure 10.3.1 Excavation. Geometry and Discretizations.

the horizontal displacements is shown in figure (10.3.3). As can be obtained from these curves, the last cohesion value for which stationary conditions are possible is $c = 0.3$ tf/m^2. This result agrees well with analytical stability procedures (Swedish circle or logspiral mechanism), which showed factor of safety near the unit when the cohesion and friction angle are chosen 0.30 tf/m^2 and 20° respectively.

As already shown in the first example, only the viscoplastic displacements are in the direction of the load, due to possible relative value nature attributed to them. The elastic displacements are in a downward direction while the load corresponding to the relief of the initial stress conditions are in the opposite direction.

(b)   Embankment

This example consists of analysing a hypothetical embankment 10 m high with a slope of 2 : 1 as in the excavation problem. The embankment is supposed to be constructed on a 5 m strata of soil which overlays bedrock. The domain defined to solve this problem is closed and contains both the embankment and foundation, which are assumed to be constituted of the same material (fig. 10.3 4).

The load applied to the structural system defined above is only due to the unit weight of the soil and also taken equal to 2 tf/m^3. All the material properties used for this application are assumed to be those adopted in the previous case.

As in the first example of chapter 5, the integrations over the domain in order to compute the self-weight effects were avoided by adopting equivalent boundary integrals.

In figures (10.3.5) and (10.3.6) the values of the horizontal and vertical displacements are presented as functions of the cohesion.

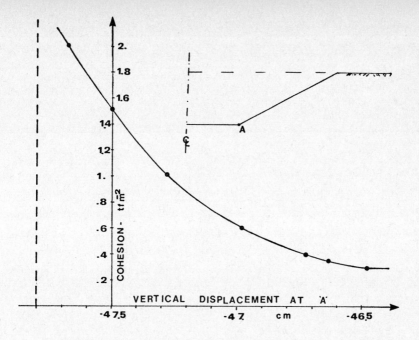

Figure 10.3.2 Vertical Displacement at Node "A".

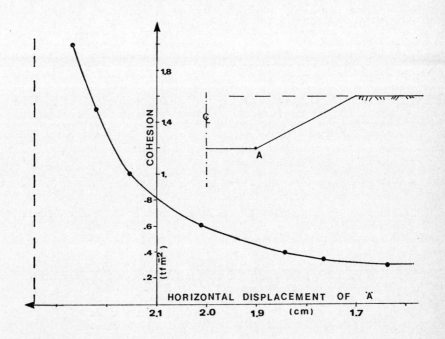

Figure 10.3.3 Horizontal Displacement at Node "A".

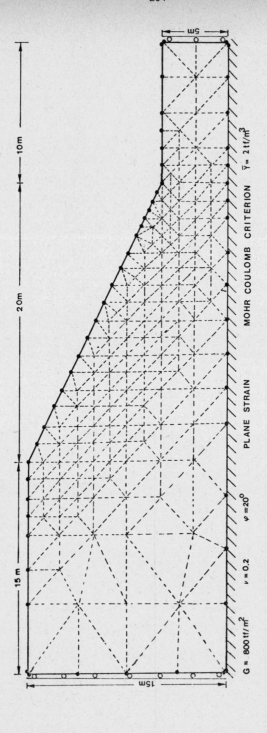

Figure 8.3.4 Embankment. Discretizations.

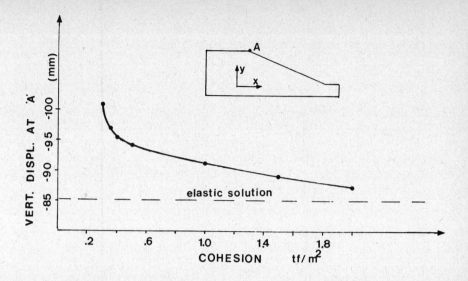

Figure 8.3.5 Vertical Displacemenr at Node "A".

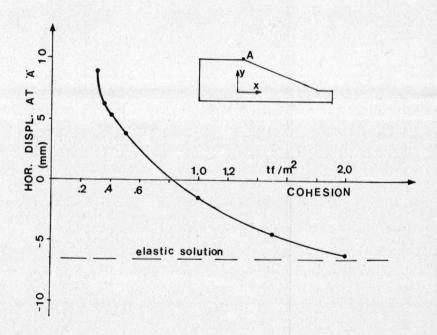

Figure 8.3.6 Horizontal Displacement at Node "A".

As in the excavation case the minimum cohesion value for which stationary conditions have been reached is equal to 0.3 tf/m^2.

The comparison between the elastic ground displacements with the corresponding values computed assuming viscoplastic behaviour and cohesion equal to 0.3 tf/m^2 is given in figure (10.3.7). These results also compare well with the finite element solution (ref. 17) obtained using the same domain and similar boundary conditions.

10.4 Tunnelling Stress Analysis

One of the common problems in rock mechanics is the study of tunnel excavations associated with the insertion of lining supports. In the early sixties some closed form solutions for circular opening were proposed for the determination of the stress distribution in the rock taking into account the plastic behaviour of the material (119, 120). Only after the development of numerical techniques could tunnelling analysis be made within a realistic context.

In tunnelling problems, both the face advance and the lining insertion have been the subject of interesting experiments and the theoretical essays of various researchers (86, 121, 122, 123, 124). The face advance is actually a three-dimensional problem; however, for the cavity stability analysis, some works have already been carried out which demonstrate the possibility of solving the problem within the context of plane analysis. The displacements perpendicular to the tunnel axis developed near the excavation face and before the lining insertion can also be computed by assuming plane conditions without representative loss of accuracy in the final results.

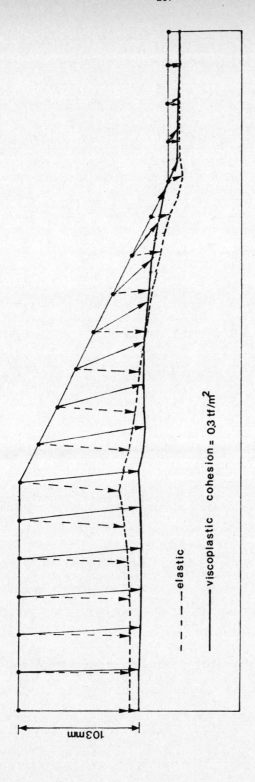

Figure 10.3.7 Embankment Surface Displacements.

The insertion of lining supports in a rock opening constitutes another important problem in any tunnel construction. The assumptions that plastic and elastic deformations occur instantaneously either before or after the lining insertion were often employed to analyse displacement and stress distribution around tunnels and in the lining support. However, more realistic concepts are now possible, and all stages of a tunnel construction can be modelled taken into acount the real time-dependent characteristics of the materials.

The example presented in this section consists of analysing the displacements around a circular opening and in the concrete lining which was inserted some time after the excavation. This problem has been taken from reference (17) where the corresponding finite element solution is presented.

For all cases analysed, the rock material was assumed to have a constant residual stress field before the tunnel is cut. The vertical stresses have been taken equal to 80000 lb/ft^2, while the horizontal stresses were computed according to the earth pressure coefficient at-rest, K_o, assumed equal to 0.5. For the rock and concrete, the following values of the elastic parameter have been adopted.

material \ parameter	rock	concrete
elastic modulus (lb/ft^2)	7.2×10^7	4.32×10^8
Poisson's ratio	0.15	0.20

Table 10.4.1 Circular tunnel. Elastic values.

Before the lining insertion the opening has an internal diameter equal to 11 ft. After the application of the concrete support the internal diameter is reduced to 10 ft. The boundary and internal discretizations, together with the geometry of the problem, are presented in figure (10.4.1). As can be seen, only a quarter of the domain needed to be discretized due to the symmetry. Several loadings are here assumed to represent different stages of the tunnel construction and they will be illustrated separately in this section.

(a) Unlined Case

This is a situation which often occurs in practical tunnelling problems, either due to the deliberate omission of the structural lining or due to a prolonged delay before the insertion of the support.

Depending on both the viscous properties of the rock and the speed of the excavation, elastoplastic or visco/elastoplastic solutions are acceptable to model the stress and displacement distribution around the cavity. Elastoplastic behaviour is assumed when the time-dependent displacements occur very quickly in comparison with the face advancing. However, the case showing long-time displacements in which inelastic deformations are noticed far behind the face and for quite a long time is the most common situation.

In order to solve the tunnel excavation assuming elastoplastic and visco/elastoplastic behaviour, Perzyna's model was employed together with an associative Mohr-Coulomb flow rule. The linear function Φ (see equation 9.3.5) was applied. The time was defined in units of $1/\gamma$ (γ is the viscosity parameter) and the incremental stepping procedure has been taken according to the criterion given in equation (9.5.3). The cohesion and the friction angle were taken to be 1.44×10^4 lb/ft^2

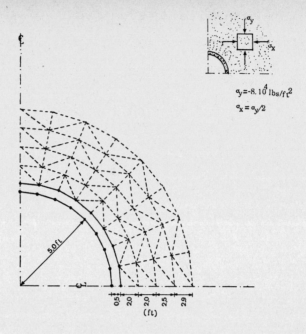

Figure 10.4.1 Circular Tunnel Discretizations.

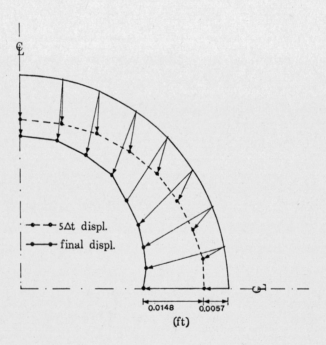

Figure 10.4.2 Displacements of the Rock Without Lining.

and $30°$ respectively. At this stage, the only load applied is due to the removal of the residual stress on the opening surface.

Assuming that the load is applied in one single increment, the final elasto/viscoplastic solution is obtained and the displacements over the tunnel surface are shown in figure (10.4.2), in which displacements after time equivalent to $5\Delta t$ are also presented.

As has already been shown in the first example of chapter 9, the plastic solution, here obtained by applying the load in increments followed by stationary conditions, does not differ significantly from the results of figure (10.4.2).

(b) Lined Case

In this case, the insertion of the lining is considered in order to restrain the viscous deformations. The loads, equivalent to the removal of the core, are applied in one increment and then the lining is placed at a practical distance from the face. As time and viscous parameter have been hypothetically assumed to solve this example, the distance from the application of the lining to the face is considered to be equivalent to a time interval equal to $5\Delta t$. Therefore, part of the viscous displacements has already occurred before the application of the support (see fig. 10.4.2).

The lining is assumed to be of concrete material and will be displaced from the initial position only due to the viscous deformation of the rock. Figure (10.4.3) shows the final steady displacements for the lining, which compare well with the original finite element solution given in reference (17).

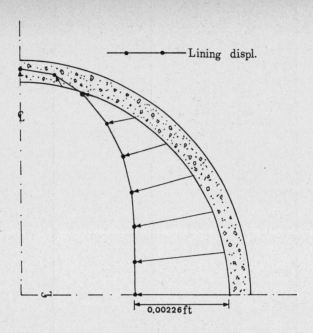

Figure 10.4.3 Final Lining Displacements.

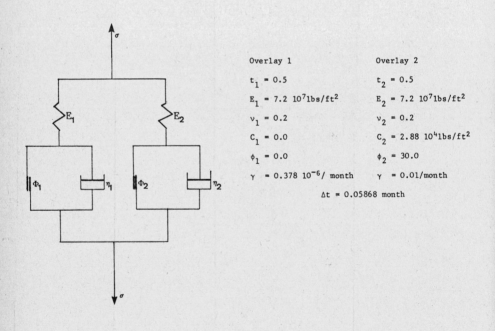

Overlay 1

$t_1 = 0.5$
$E_1 = 7.2 \; 10^7 \text{lbs/ft}^2$
$\nu_1 = 0.2$
$C_1 = 0.0$
$\phi_1 = 0.0$
$\gamma = 0.378 \; 10^{-6}/\text{month}$

Overlay 2

$t_2 = 0.5$
$E_2 = 7.2 \; 10^7 \text{lbs/ft}^2$
$\nu_2 = 0.2$
$C_2 = 2.88 \; 10^4 \text{lbs/ft}^2$
$\phi_2 = 30.0$
$\gamma = 0.01/\text{month}$

$\Delta t = 0.05868$ month

Figure 10.4.4 Rheological Model.

(c) Alternative Case

The transference of the stress from the rock to the lining support obtained in the last example is only due to the time-dependent deformations which took place in a small region (viscoplastic zone) in the vicinity of the interface between concrete and rock materials. However, in many tunnel designs it is convenient to assume that viscoelastic or creep effects occur for any change in the rock stress level. Using the overlay concept we can define a model to give both viscoplastic and viscoelastic responses. Here, the viscoplastic effects are modelled assuming the same conditions adopted in cases "a" and "b", and the viscoelastic deformations are governed by an equivalent Kelvin-Voight unit. Figure (10.4.4) shows the rheological model obtained by representing the rock material using two overlay models. Also in figure (10.4.4) the parameters adopted to simulate the viscoelastic and viscoplastic behaviours are presented. The internal discretization for this example has to be sufficiently increased in order to take into account all significant viscous deformations.

As was expected with the assumptions of viscous effects everywhere, considerable changes in the final solution were verified. Figure (10.4.5) presents the boundary displacements after five time steps which also represent the time of the lining application. In this case, the displacements due to viscous deformations occur faster than in the previous case. As a consequence, the final steady solution (fig. 10.4.5) shows smaller displacements in the lining, and the final forces acting on the interface between concrete and rock are smaller in comparison with the corresponding values obtained in "b" (fig. 10.4.6).

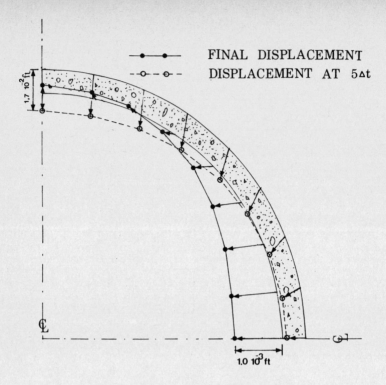

Figure 10.4.5 Rock Displacements After 5Δt and Final Lining Displacements.

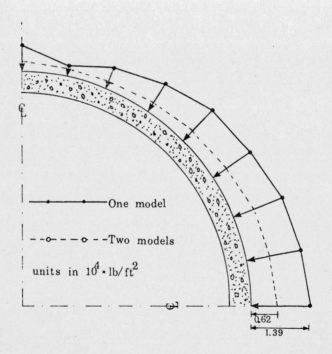

Figure 10.4.6 Reaction on the Lining Surface.

Another comparison between the cases presented is given by the normal stresses on the interface taken in the tangential direction, as can be seen in figure (10.4.7). These results illustrate the effects which are obtained in the rock material as a function of both the lining insertion time and the viscous properties.

(d) Operation Tunnel

One of the uses of a lined tunnel as in "a", "b" and "c" is to carry water under high pressure conditions. In order to simulate the operation conditions encountered by a tunnel, the application of a high internal pressure will be considered in this example.

Two possibilities are now examined and in both cases stationary conditions have been considered achieved for the excavation effects. In the first case, the excavation effects are supposed to reach stationary conditions before the insertion of the support; as a consequence, the lining does not show any initial prestressed state. In the second case, the same internal load is applied, but now assuming that the lining has been previously compressed due to the viscous effects, as in case "b".

The internal load chosen to run this example is equal to 1.5×10^5 lb/ft^2, and is assumed to be applied in one increment only. Figure (10.4.8) shows the final lining displacements in both cases in comparison with the previous values presented in figure (10.4.3). The normal stress values in the direction tangential to the interface are also illustrated for both cases in comparison with those resulting from the previous analyses without internal pressure (figs. 10.4.9 and 10.4.10). As can be seen, the viscous effect consideration can reduce considerably the final lining stress distribution.

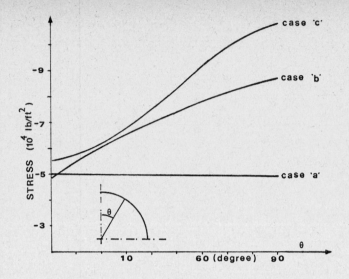

Figure 10.4.7 Normal Rock Stress Distribution on the Interface.

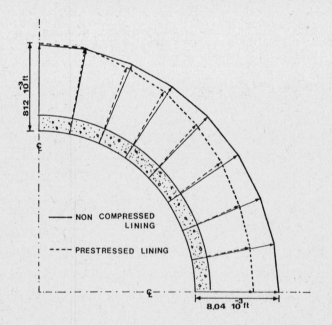

Figure 10.4.8 Final Lining Displacements After Applying the Internal Load.

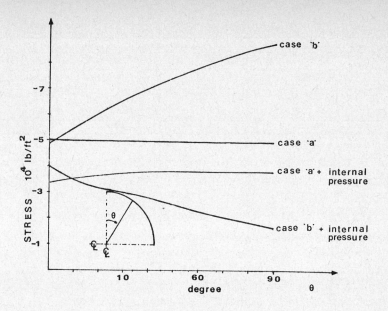

Figure 10.4.9 Rock Normal Stresses Along the Interface.

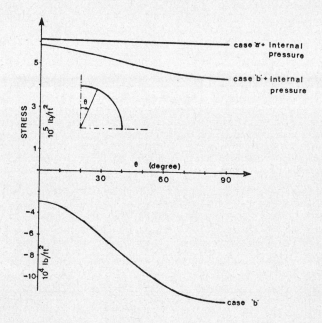

Figure 10.4.10 Lining Normal Stresses Along the Interface.

CHAPTER 11

CONCLUSIONS

The basic objective of this work was to develop numerical procedures for the boundary element method to analyse geotechnical problems of practical interest. Linear and nonlinear formulations such as no-tension, plasticity and others have been implemented to deal with a wide range of two-dimensional problems.

In chapter 3 the complete formulation of the integral equations for plane stress and plane strain problems has been presented. The proper way to consider deformation type load such as temperature, shrinkage, swelling and others was developed. Simple expressions were also formulated for computing self-weight and prestress forces often necessary in geotechnical problems.

The complete plane strain idealization was formulated in chapter 4 mainly to model long openings in which the axes are not coincident with a principal direction. As expected, two uncoupled problems, corresponding to the plane strain and anti-plane cases, were obtained. The first problem was already formulated in chapter 3, therefore only the integral equations related to the anti-plane case have been developed. In common with the plane case, the proper procedure for obtaining the complete expressions for stresses at internal points were formulated taking into account initial stress and strain type loads.

The integral equations for plane and anti-plane problems presented in chapters 3 and 4 are shown transformed into what is known as the Boundary Element Method, after performing the necessary boundary and domain integrals. These integrals are computed employing either analytical or numerical schemes. For some examples presented in this work, the Gaussian quadrature scheme (24) has proved to give accurate results using only six integration points. No significant change in the results took place when more interpolation points were used for boundary and domain integrals.

In order to save computer time a scheme has been adopted in which the number of integration points is chosen to be a function of the distance between the load point and either the boundary element or the internal cell under consideration. In the case of boundary integrals, the scheme adopted to choose the number of integration points is given in table (11.1), where ℓ is the size of the element and d is the distance from the load point to the element middle point.

distance	number of integration points
$d < 2\ell$	6
$2\ell \leq d < 6\ell$	4
$6\ell \leq d$	2

Table 11.1 Number of integration points for elements

It is important to notice that the boundary integral accuracy is also dependent upon the orientation of the element in relation to the load point, a factor which is not taken into account in the scheme presented.

For the integration over internal cells, a similar scheme was adopted. This time the number of integration points was chosen to be governed by the maximum angle defined by the cell and load point, as given in table (11.2).

angle	number of integration points
$\alpha \geq 60$	6
$60 > \alpha \geq 15$	4
$15 > \alpha$	2

Table 11.2 Number of integration points for cells.

The applicability of the complete plane strain and the thin subregion formulations in conjunction with the boundary element method were illustrated by the results obtained from two tunnel analyses presented in chapter 5. In the same chapter, the uses of both the boundary integrals for computing body forces effects and the subregion technique for taking into account domains formed by different materials were also illustrated with practical examples for which accurate results were obtained.

Although the lined tunnel analysis (subregion problem) gave accurate results, it has been found during the implementation of the computer program that the subregion technique often requires fine boundary discretizations to improve the quality of the numerical solutions. As a consequence, the large system of equations formed increases the computer time. Thus subregions are to be avoided wherever possible. The time saving characteristically attributed to the subregions due to both the formation of banded matrices and the small number of integrals vanishes if a proper reduction scheme of the integration point number were adopted together with the use of a single region.

The extension of the boundary element technique to no-tension problems has been presented in chapter 6. The results obtained for practical applications in rock mechanics illustrate the efficiency and accuracy of the approach. The solution technique employed to compute the initial stress effects, together with the proper simulation of the infinite domain, are the advantages of the formulation by comparison with other no-tension approaches.

Two criteria related to the unloading path of the stress-strain curve have been proposed. The first assumes the total recovery of the tension deformation after the removal of the load; in such instances no incremental procedure has to be applied and the solution is obtained quickly. In the second case however, an incremental scheme must be adopted in order to follow the load path.

Situations in which small rock tensile strengths are present have not been analysed, however they can be modelled following a similar procedure. For this case, the approach must take into account the cracks formed in order to eliminate any tensile stresses which might occur for further increments of load. It should be noticed that for this analysis the approach must be incremental.

The second nonlinear behaviour adapted to boundary element techniques is the modelling of rock discontinuities. The crack directions have to be defined by the discretization and conditions to govern displacements and traction over the crack surface are introduced. The solution obtained with this technique showed the reliability of the approach in practical situations, although the process can be slow in some cases due to the necessity of solving the system of equations many times. For instance, analysis of blocked rocks requires the

definition of small subregions separated by interfaces over which slip
or separation might occur. Therefore, in this case a larger system
of equations is obtained, resulting in increase of computer time.

In chapters 8 and 9 the elastoplastic and elasto/viscoplastic
algorithm for boundary element formulation were also proposed employing
the solution technique presented in chapter 5. Both approaches have been
verified to give accurate results in comparison with some finite element
solutions. As has been discussed for no-tension, the need for cell
discretization over only the parts of the domain where the inelastic
deformation might occur is another reason why computer time is saved
and the data requirement of the problem is reduced.

The semi-analytical procedure adopted in chapter 5 to evaluate the
effects of an initial stress field applied in the domain, together with
the solution technique and the scheme for the reduction of the number
of integration points, have proved to be accurate and efficient for
the examples solved. In these applications, plastic and viscoplastic
solutions have been obtained with relatively coarse boundary and internal
discretizations when compared with the corresponding finite element
grids employed to solve the same examples.

The example discussed in chapters 8 and 9 also showed that the
viscoplastic algorithm (used to model plastic behaviour) is quicker than
the plastic approach when the same incremental procedure is adopted
for both, and the time steps are taken to be equal to the limit given
in chapter 9. However, for the examples presented and many other tests
carried out during this work, inaccuracies generally smaller than 0.5%
were verified when the viscoplastic algorithm was used together with
the Cormeau's time step limit, although for geotechnical engineeering

such inaccuracies are acceptable and easily improved when smaller values of the time step are adopted. The maximum errors decrease to values of around 0.01% when the time step is reduced by a factor of 5.

The viscoplastic formulation, in conjunction with overlay technique for two examples discussed in chapter 9, demonstrate the efficiency of boundary elements in dealing with more complex nonlinear behaviours. In these cases, boundary and finite elements have produced almost the same solution despite employing different procedures to compute the inelastic deformations.

The potential of any algorithm or method must be proved when used to model collapse loads and the mechanism of rupture. In chapter 10 some recognized difficult problems in geomechanics have been analysed with the algorithms proposed in this work. The first example presents the study of the collapse load for a strip footing. It is well known that the analysis of this problem using finite elements can only achieve reasonable results if fine meshes are adopted. When coarse meshes are employed the representation of the limit load is impossible due to constraints created by the incompressibility of the plastic strain (125). In finite element analysis the inability to represent limit loads is found when the internal strains are computed from the derivatives of interpolated displacements, and this can be avoided by employing convenient approximations (54, 55).

Using the boundary element formulation a similar problem was verified for the strip footing example. It was solved only by increasing the discretization on the boundary alone; this led to a limit load only 1% above the predicted value. It is worth mentioning that in the analysis carried out using subregions, the problem of predicting a limit load becomes more difficult, and very fine discretizations on the interface were needed.

The second limit analysis problem studied in chapter 10, the slope stability case, is regarded as a difficult task for any numerical technique. As can be seen by the meshes used to solve this example, fine boundary discretizations were necessary to model accurately the system collapse.

The final application concerned the use of the viscoplastic algorithm to model actual time-dependent responses. The procedure adopted to analyse stress distribution in an underground opening is shown. It illustrates that the viscous deformations can be used to improve the final stress conditions in the lining, if the time when lining is constructed is adequately selected.

In all examples presented in this work the quality of the results obtained has been pointed out demonstrating the potential of the approaches formulated; however, no reference has been made to the computer time spent for running practical cases. Although no comparison between the time spent for running the presented problems using boundary elements and other numerical techniques has been carried out using the same computer, it is worth mentioning the time required to solve some of the examples. Taking into account that the algorithm employed for the iterative process is simple and economic due to the inelastic effects being computed by a simple matrix-vector product, the main doubt with regard to efficiency is the time spent for assembling the matrices needed to solve the problem. Thus the time consumed in order to compute the matrices (inclusive of the inversion of $\underline{A}$) for some of the examples presented is given in table (11.3).

Problem	ICL 2970 time (sec.)
strip footing one subregion	1096
strip footing two subregions	1632
excavation	1182

Table 11.3 Time spent to assemble the matrices

Notice that for these problems the necessary domain and boundary integrations are performed over numbers of cells and elements which are twice those shown by the meshes due to the consideration of the axis of symmetry.

The total time (assembling of matrices plus iterative procedure) is given for the tunnel problem in table (11.4). Notice that the excessive time consumed by case "c" is due to the large discretization adopted in order to compute the viscoelastic responses.

Problem	ICL 2970 time (sec)	time steps
unlined case - case "a"	237	177
lining displacements - case "b"	218	31
lining displacements - case "c"	992	241

Table 11.4 Total time spent for tunnelling analysis.

Several improvements have been introduced into the direct boundary element method, mainly concerned with the application of the technique to solve geomechanical problems. Many practical problems have been analysed with the different approaches formulated, and the results obtained show the applicability of the method to a wide range of situations.

REFERENCES

1. BREBBIA, C.A., "The Boundary Element Method for Engineers", Pentech Press, London, 1978.

2. BREBBIA, C.A. and WALKER, S., "Boundary Element Techniques in Engineering", Newnes-Butterworths, London, 1980.

3. BREBBIA, C.A., TELLES, J.C. and WROBEL, L.C., "Boundary Elements. Fundamentals and Applications in Engineering", Spring-Verlag, 1983. (in press).

4. RIZZO, F.J., "An Integral Equation Approach to Boundary Value Problems of Classical Elastostatics", Quart. Appl. Math., 25, pp. 83-95, 1967.

5. CRUSE, T.A., "Mathematical Foundations of the Boundary Integral Equation Method in Solid Mechanics", Report No. AFOSR-IR-77-1002, Pratt and Whitney Aircraft Group, 1977.

6. JASWON, M.A. and SYMM, S.T., "Integral Equation Method in Potential Theory and Elastostatics", Academic Press, London, 1977.

7. FINLAYSON, B.A., "The Method of Weighted Residual and Variational Principles", Academic Press, 1972.

8. KUPRADZE, V.D., "Potential Methods in Theory of Elasticity", Israel Program for Scientific Translations, Jerusalem, 1965.

9. WATSON, J.O., "On the Integral Representation of the Displacement of an Elastic Body", Report CE/19/1968, University of Southampton, 1968.

10. DRUCKER, D.C., "Limit Analysis of Two and Three Dimensional Solid Mechanics Problems", J. Mech. Phys. Solids, 1, pp. 217-226, 1953.

11. DRUCKER, D.C., "Coulomb Friction, Plasticity and Limit Loads", J. Appl. Mech., 21, pp. 71-74, 1954.

12. DRUCKER, D.C. and PRAGER, W., "Soil Mechanics and Plastic Analysis or Limit Design", Quart. Appl. Math., 10, pp. 157-165, 1952.

13. ROSCOE, K.H. and BURLAND, J.B., "On the Generalised Stress-Strain Behaviour of 'Wet' Clay", in Engineering Plasticity (ed. by J. Heyman and F.A. Leckie), University Press, Cambridge, pp. 535-609, 1968.

14. LAMA, R.D. and VUTUKURI, V.S., "Handbook on Mechanical Properties of Rocks", 3, Trans. Tech. Publications, 1978.

15. GRIBBS, D., "Creep of Rocks", J. Geol., 47, pp 225-251, April - May 1939.

16. PERZYNA, P., "Fundamental Problems in Viscoplasticity", Advances in Applied Mechanics, 9, pp. 243-377, Academic Press, New York, 1966.

17. HUMPHESON, C., "Finite Element Analysis of Elasto/Viscoplastic Soils", Ph.D. Thesis, University of Wales, Swansea, 1976.

18. CORMEAU, I.C., "Viscoplasticity and Plasticity in the Finite Element Method", Ph.D. thesis, University of Wales, Swansea, 1976.

19. SOUTHWELL, R.V., "Relaxation Methods in Theoretical Physics", Oxford University Press, London, 1946.

20. SOARE, M., "Application of Finite Difference Equations to Shell Analysis", Pergamon Press, 1967.

21. TURNER, M.J., CLOUGH, R.W., MARTIN, H.C. and TOPP, L.J., "Stiffness and Deflection Analysis of Complex Structures", J. Aero Sci., 23, pp. 805-823, 1956.

22. CLOUGH, R.W., "The Finite Element in Plane Stress Analysis", Proc. 2nd A.S.C.E. Conf. on Electronic Computation, Pittsburg, 1960.

23. HAMMER, P.C., MARLOVE, O.J. and STROUD, A.H., "Numerical Integration over Simplex and Cones", Mathematics of Computation, 10, 1956.

24. STROUD, A.H. and SECREST, D., "Gaussian Quadrature Formulas", Prentice Hall, New York, 1966.

25. WATSON, J.O., "The Analysis of Thick Shells with Holes, by Integral Representation of Displacement", Ph.D. Thesis, University of Southampton, 1972.

26. WATSON, J.O., "The Analysis of Three-Dimensional Problems of Elasticity by Integral Representation of Displacement", Proc. of the Int. Conf. on Variational Methods in Engineering, University of Southampton, 1972.

27. CRUSE, T.A., "Numerical Solutions in Three-Dimensional Elastostatics", Int. J. Solids Struct., 5, pp. 1259-1274, 1969.

28. LOVE, A.E.H., "Treatise on the Mathematical Theory of Elasticity", Dover, 1944.

29. NAKAGUMA, R.K., "Three-Dimensional Elastostatics Using the Boundary Element Method", Ph.D. thesis, University of Southampton, 1979.

30. MINDLIN, R.D., "Force at a Point in the Interior of a Semi-Infinite Solid", J. Physics, 7, 1936.

31. TELLES, J.C.F. and BREBBIA, C.A., "Boundary Element Solution for Half-Plane Problems", Int. J. Solids Struct., 17, pp. 1149-1158, 1981.

32. MELAN, E., "Der Spannungszustand der Durch eine Einzelkraft im Innern Beanspruchten Halbscheibe", Z. Angew, Math. Mech., 12, pp. 1251-1267, 1932.

33. MELNIKOV, Y.A., "Surface Integral Method Applied to the Problems of an Elastic Strip with Periodically Spaced Holes", J. Appl. Mech., pp. 599-607, 1976.

34. RICARDELLA, P.C., "An Implementation of the Boundary Integral Technique for Planar Problems in Elasticity and Elastoplasticity", Report. No. SM-73-10, Dept. Mech. Engng., Carnegie Mellon University, Pittsburg, 1973.

35. MENDELSON, A., "Boundary Integral Methods in Elasticity and Plasticity", Report No. NASA TN D-7418, NASA, 1973.

36. MUKHERJEE, S., "Corrected Boundary Integral Equation in Planar Thermoelastoplasticity", Int. J. Solids Struct., 13, pp. 331-335, 1977.

37. CHAUDONNERET, M., "Méthode des Équations Intégrales Appliquées a la Résolution de Problêmes de Viscoplasticité," J. Méchanique Appliquée, 1, pp. 113-132, 1977.

38. TELLES, J.C.F. and BREBBIA, C.A., "On the Application of the Boundary Element Method to Plasticity", Appl. Math. Modelling, 3, 466-470, 1980.

39. TELLES, J.C.F. and BREBBIA, C.A., "Elastoplastic Boundary Element Analysis", Proc. Europe - U.S. Workshop on Nonlinear Finite Element Analysis in Structural Mechanics, (Wunderlich et al., eds.), Ruhr-University Bochum, Germany, pp. 403-434, 1980.

40. TELLES, J.C.F. and BREBBIA, C.A., "Elastic/Viscoplastic Problems Using Boundary Elements", Int. J. Mech. Sci., in press.

41. MENDELSON, A., "Plasticity: Theory and Application", Macmillan, New York, 1968.

42. ZIENKIEWICZ, O.C., VALLIAPAN, S. and KING, I.P., "Elasto-Plastic Solutions of Engineering Problems, Initial Stress Finite Element Approach", Int. J. Num. Meth. Engng., 1, pp. 75-100, 1969.

43. MIKHLIN, S.G., "Singular Integral Equation", Amer. Math. Soc. Trans. Series 1, 10, pp. 84-197, 1962.

44. SOMIGLIANA, C., "Sopra l'Equilibrio di un Corpo Elastico Isotropo", Il Nuovo Ciemento, t. 17-19, 1886.

45. BETTI, E., "Teoria dell Elasticita", Il Nuovo Ciemento, t. 7-10, 1872.

46. VALLIAPPAN, S., "Non-Linear Stress Analysis of Two-Dimensional Problems with Special Reference to Rock and Soil Mechanics", Ph.D. thesis, University of Wales, Swansea, 1968.

47. ZIENKIEWICZ, O.C., NAYAK, G.C. and OWEN D.R.J., "Composite and Overlay Models in Numerical Analysis of Elasto-Plastic Continua", Int. Symp. Foundations of Plasticity, Warsaw, 1972.

48. OWEN, D.R.J., PRAKASH, A. and ZIENKIEWICZ, O.C., "Finite Element Analysis of Non-Linear Composite Materials by Use of Overlay Systems", Computers and Structures, 4, pp. 1251-1267, 1974.

49. PANDE, G.N., OWEN, D.R.J. and ZIENKIEWICZ, O.C., "Overlay Models in Time-Dependent Nonlinear Material Analysis", Computers and Structures, 7, pp. 435-443, 1977.

50. HILL, R., "The Mathematical Theory of Plasticity", Clarendon Press, Oxford, 1950.

51. PRANDTL, L., "Über die Eindringungsfestigkeit (Härte) Plastische Baustoffe und die Festigkeit von Schneiden", Z. Angew. Math. Mech., 1, pp. 15-20, 1921.

52. TERZAGHI, K., "Theoretical Soil Mechanics", Wiley, New York, 1943.

53. CHEN, W.F., "Limit Analysis and Soil Plasticity", Elsevier Scientific Publishing Company, Amsterdam, 1975.

54. BÄTHE, K.J., SNYDER, M.D. and CIMENTO, A.P., "On Finite Element Analysis of Elasto-Plastic Response", Int. Conf. on Eng. Application of the Finite Element Method, Høvik, Norway, 1979.

55. SLOAN, S.W. and HANDOLPH, M.F., "Numerical Prediction of Collapse Loads Using Finite Element Methods". Int. J. Num. An. Meth. Geom., 6, pp. 47-76, 1982.

56. WU, R.S. and CHOU, Y.T., "Line Force in a Two-Phase Orthotropic Medium", J. Appl. Mech., 49, pp. 55-61, 1982.

57. NAYAK, G.C. and ZIENKIEWICZ, O.C., "Convenient form of Stress Invariants for Plasticity", Proc.Am. Soc. Civ. Engrs., J. Struct. Div., 98, pp. 949-954, 1972.

58. CLOUGH, R.W. and RASHID, Y., "Finite Element Analysis of Axisymmetric Solids", ASCE, J. Eng. Mech. Div., 91, N.EM1, pp. 71-85, 1965.

59. HOCKING, G., "Three Dimensional Elastic Stress Distribution around the Flat End of a Cylindrical Cavity", Int. J. Rock Mech. Min. Sci., 13, pp. 331-337, 1976.

60. LACHAT, J.C. and WATSON, J.O.A., "A Second Generation Boundary Integral Equation Program for Three Dimensional Elastic Analysis", in Boundary Integral Equation Method : Computational Applications in Applied Mechanics, ed. by T.A. Cruse and F.J. Rizzo, ASME, pp. 85-100, 1975.

61. BRADY, B.H.G and BRAY, J.W., "The Boundary Element Method for Determining Stress and Displacements around Long Openings in a Triaxial Stress Field", Int. J. Rock Mech. Min. Sci.,15, pp. 21-28, 1978.

62. BRADY, B.H.G., "A Direct Formulation of the Boundary Element Method of Stress Analysis for Complete Plane Strain", Int. J. Rock Mech. Min. Sci., 16, pp. 235-244, 1979.

63. JAEGER, J.C. and COOK, N.G.W., "Fundamentals of Rock Mechanics", John Wiley and Sons, Inc. New York, 1976.

64. BRADY, B.H.G. and BRAY J.W., "The Boundary Element Method for Elastic Analysis of Tabular Orebody Extraction, Assuming Complete Plane Strain", Int. J. Rock Mech. Min Sci., 15, pp. 29-37, 1978.

65. HOCKING, G., "Stress Analysis of Underground Excavations Incorporating Slip and Separation Along Discontinuities", in Recent Advances in Boundary Element Methods, ed. by C.A. Brebbia, Pentech Press, 1978.

66. WARDLE, L.J. and CROTTY, J.M., "Two-Dimensional Boundary Integral Equation Analysis for Non-Homogeneous Mining Applications", in Recent Advances in Boundary Element Methods, ed. by C.A. Brebbia, Pentech Press, 1978.

67. CHAUDONNERET, M., "Resolution of Tension Discontinuity Problems in Boundary Integral-Equation Method Applied to Stress Analysis", Comptes Rendus, Academie des Sciences, Ser. A., Math, 284, pp. 463-466, 1977.

68. BRADY, B.H.G. and WASSYNG, A., "A Coupled Finite Element - Boundary Element Method of Stress Analysis", Int. J. Rock Mech. Min. Sci., 18, pp.475-485, 1981.

69. DENDROU, B.A., and DENDROU, S.A., "A Finite Element Boundary Integral Scheme to Simulate Rock Effects on the Liner of an Underground Intersection", in Boundary Element Methods, ed. by C.A. Brebbia, Springer-Verlag, 1981.

79. NAYLOR, D.J. and ZIENKIEWICZ, O.C., "Settlement Analyses of a Strip Footing Using a Critical State Model in Conjunction with Finite Elements", Proc. Symp. Interaction of Structure and Foundation, Birmingham, 1972.

80. HÖEG, K., "Finite Element Analysis of Strain-Softening Behaviour", Proc. Int. Conf. Comput. Meth. Nonlinear Mech., Austin, Texas, pp. 969-978, 1974.

81. MENDELSON, A. and ALBERS, L.U., "Application of Boundary Integral Equation to Elastoplastic Problems", in Boundary Integral Equation Method : Computational Application in Applied Mechanics, ed. by T.A. Cruse and J.F. Rizzo, ASME, pp. 47-84, 1975.

82. VENTURINI, W.S. and BREBBIA, C.A., "The Boundary Element Method for the Solution of No-Tension Materials", in Boundary Element Methods, ed. by C.A. Brebbia, Springer-Verlag, 1981.

83. VENTURINI, W.S. and BREBBIA, C.A., "Boundary Element Formulation to Solve No-Tension Problems in Geomechanics", in Numerical Method in Geomechanics, ed. by J.B. Martins, 1982.

84. VENTURINI, W.S. and BREBBIA, C.A., "Some Applications of Boundary Element Methods in Geomechanics", Int. J. Num. Anal. Math. Geom. (in press).

85. ZIENKIEWICZ, O.C., WATSON, M. and KING, I.P., "A Numerical Method of Visco-Elastic Stress Analysis", Int. J. Mech. Sci., 10, pp. 807-827, 1968.

86. LOMBARDI, G., "Long Term Measurements in Underground Opening and their Interpretation with Special Consideration to the Rheological Behaviour of the Rock", in Field Measurements in Rock Mechanics, ed. by K. Kovari, 2, pp. 839-858, 1977.

70. CLOUGH, R.W. and WOODWARD, R.J., "Analysis of Embankment Stresses and Deformations", J. Soil Mech. Found. Div., ASCE. 93, N. SM4, pp. 529-549, 1967.

71. DUNLOP, P. and DUNCAN, J.M., "Development of Failure around Excavated Slopes", ASCE, J. Soil Mech. Found Div., 96, N. SM2, pp. 471-493, 1970.

72. DESAI, C.S. and REESE, L.C., "Analysis of Circular Footings on Layered Soils", J. Soil Mech. Found. Div., ASCE, 96, N. SM4, pp. 1289-1310, 1970.

73. DUNCAN, J.M. and CHANG, C.Y., "Non-Linear Analysis of Stress and Strain in Soil", J. Soil. Mech. Found. Div., ASCE, 96, N. SM5, pp. 1629-1653, 1970.

74. CHANG, C.Y. and DUNCAN, J.M., "Analysis of Soil Movement around a Deep Excavation", J. Soil Mech. Found. Div., ASCE, 96, N. SM5, pp. 1655-1681, 1970.

75. GOODMAN, R.E., TAYLOR, R.L. and BREKKE, T.L., "A Model for the Mechanics of Jointed Rock", J. Soil Mech. Found. Div., ASCE, 94, pp. 637-657, 1968.

76. REYES, S.F. and DEERE, D.U., "Elasto-Plastic Analysis of Underground Openings by the Finite Element Method", Proc. 1st Congress of the Int. Soc. of Rock Mech., 2, pp. 477-483, 1966.

77. ROSCOE, K.H., SCHOFIELD, A.N. and WORTH, C.P., "On the Yielding of Soils", Geotechnique, 8, pp. 25-53, 1958.

78. ROWE, P.W., "The Stress-Dilactancy Relation for Static Equilibrium of an Assembly of Particles in Contact", Proc. R. Soc. A., 269, pp. 500-517, 1962.

87. VENTURINI, W.S. and BREBBIA, C.A., "Boundary Element Formulation for Nonlinear Applications in Geomechanics", App. Math. Modelling (in press).

88. MUKHERJEE, S. and KUMAR, V., "Numerical Analysis of Time-Dependent Inelastic Deformation in Metallic Media Using the Boundary Integral Equation Method", Trans. ASME, J. Appl. Mech., 45, pp. 785-790, 1978.

89. MORJARIA, M. and MUKHERJEE, S. "Improved Boundary Integral Equation Method for Time-Dependent Inelastic Deformation in Metals", Int. J. Num. Meth. Engng, 15, pp. 97-111, 1981.

90. BUI, H.D., "Some Remarks About the Formulation of Three-Dimensional Thermoelastic Problems by Integral Equations", Int. J. Solids Structures, 14, pp. 935-939, 1978.

91. HARTMANN, F., "Elastic Potentials on Piecewise Smooth Surfaces", Journal of Elasticity, 12, pp. 31-50, 1982.

92. HARTMANN, F., "Computing the C-Matrix on Non-Smooth Boundary Points", in New Development in Boundary Element Methods, ed. by C.A. Brebbia, CML Publications, 1980.

93. RIZZO, F.J. and SHIPPY, D.J., "An Advanced Boundary Integral Equation Method for Three-Dimensional Thermoelasticity", Int. J. Num. Meth. Eng., 11, pp. 1753-1268, 1977.

94. STIPPES, M. and RIZZO, F.J., "A Note on the Body Force Integral of Classical Elastostatics", Zamp, 28, pp. 339-341, 1977.

95. DANSON, D.J., "A Boundary Element Formulation of Problems in Linear Isotropic Elasticity with Body Forces", in Boundary Element Methods, ed. by C.A. Brebbia, Springer-Verlag, 1981.

96. KELLOG, O.D., "Foundations of Potential Theory", Springer, Berlin, 1929.

97. BREBBIA, C.A. and CONNOR, J.J., "Fundamentals of Finite Element Techniques for Structural Engineers", Butterworths, 1973.

98. LACHAT, J.C., "A Further Development of the Boundary Integral Technique for Elastostatics", Ph.D. Thesis, University of Southampton, 1975.

99. BATHE, K.J., OZDEMIR, H. and WILSON, E.L., "Static and Dynamic Geometric and Material Non-Linear Analysis", Report No. UCSESM74-4, Dept. Civil.Engng., Univ. of California, Berkeley, 1974.

100. JAEGER, C., "Rock Mechanics and Engineering", Cambridge University Press, 1972.

101. ZIENKIEWICZ, O.C., TAYLOR, R.L. and PANDE, G.N., "Quasi-Plane Strain in the Analysis of Geological Problems", in Computer Methods in Tunnel Design, ICE, London, 1978.

102. CAMARGO, W.M., "Projeto de Túneis em Maciço Rochoso sob Pressão Hidrostática Interna", Ph.D. Thesis, University of Sao Paulo, 1968.

103. WITTKE, W., "New Design Concepts for Underground Opening in Rock", in Finite Elements in Geomechanics, ed. by G. Gudehus, Wiley, 1977.

104. OWEN, D.R.J. and HILTON, E., "Finite Elements in Plasticity: Theory and Practice", Pineridge Press Ltd., Swansea, 1980.

105. NAYAK, G.C., "Plasticity and Large Deformation Problems by the Finite Element Method", Ph.D. thesis, University of Wales, 1971.

106. COULOMB, C.A., "Essay sur Une Application des Regles des Maximes et Minimis a Quelques Problemes de Statique Relatifs a L'Architecture", Mem. Acad. Roy. Press Divers Sav., 5,7, Paris, 1776.

107. PERZYNA, P., "The Constitutive Equations for Rate Sensitive Plastic Materials", Quart. App. Math., 20, pp. 321-332., 1963.

108. OLSZAK, W. and PERZYNA, P.,"On Elasto-Viscoplastic Soil", in Rheology and Soil Mechanics IUTAM Symposium. Spring-Verlag, Grenoble, 1966.

109. OLSLAK, W. and PERZYNA, P.; "Stationary and Non-Stationary Visco-Plasticity", in Inelastic Behaviour of Solids, ed. by M.F. Kanninen et al., McGraw-Hill, 1970.

110. SUTHERLAND, W.H., "AXICRIP - Finite Element Computer Code for Creep Analysis of Plane Strain and Axisymmetric Bodies", Nucl. Eng. and Design, 11, pp. 269-285, 1970.

111. CORMEAU, I.C.,"Numerical Stability in Quasi-Static Elasto/ Viscoplasticity", Int. J. Num. Meth. Engng., 9, pp. 109-127, 1975.

112. ZIENKIEWICZ, O.C. and CORMEAU, I.C., "Visco-Plasticity - Plasticity and Creep in Elastic Solids - A Unified Numerical Solution Approach", Int. J. Num. Meth. Engng., 8, pp. 821-845, 1974.

113. ZIENKIEWICZ,O.C., NORRIS, V.A., WINNICKI, L.A. NAYLOR, D.J. and LEWIS, R.W., "A Unified Approach to the Soil Mechanics Problems of Offshore Foundations", in Numerical Methods in Offshore Engineering, ed. by O.C. Zienkiewicz, R.W. Lewis and K.G. Stagg, John Wiley and Sons, 1978.

114. TELLES, J.C.F. and BREBBIA, C.A., "New Developments in Elastoplastic Analysis" in Boundary Element Methods, ed. by C.A.Brebbia, Springer-Verlag, 1981.

115. SPENCER, A.J.M., "Perturbation Methods in Plasticity. III, Plane Strain of Ideal Soils and Plastic Solids with Body Forces", J. Mech. Phys. Solids, 10, pp. 165-177, 1962.

116. BESSELING, J.F., "A Theory of Elastic, Plastic and Creep Deformations of an Initially Isotropic Material Showing Anisotropic Strain-Hardening, Creep Recovery, and Secondary Creep", J. App. Mech., pp. 529-536, December, 1958.

117. LODE, W., "Versuche Ueber den Einfluss Dermit leren Hampts Pannung auf das Fliessen der Metalle Eisen Kupfen und Niekel", Z. Physik, 36, pp. 913-939, 1926.

118. FELLENIUS, W., "Calculation of the Stability of Earth Dams", Trans. 2nd Congr. Large Dams, Washington, 4, 1936.

119. KSTNER, H., "Statik de Tunnel - und Stollenbanes", Springer-Verlag, 1962.

120. TALOBRE, J.A., "La Méchanique des Roches et ses Applications", Dunod, Paris, 1967.

121. LOMBARDI, G., "Dimensioning of Tunnel Linings with Regard to Construction Procedure", Tunnels and Tunnelling, pp. 340-351, 1973.

122. LOMBARDI, G., "The Problems of Tunnel Supports", Proceedings of the Third Congress of the International Society for Rock Mechanics, VII, pp. 109-115, Denver, 1974.

123. AMBERT, W.H. and LOMBARDI, G., "Une Méthode de Calcul Élasto-Plastique de L'État de Tension et de la Deformation au tor d'une Cavité Souterraine, 3rd International Congress of Rock Mechanics, pp. 1055-1060, Denver, 1974.

124. SAKURAI, S., "Approximate Time-Dependent Analysis of Tunnel Support Structure Considering Progress of Tunnel Face", Int. J. Num. and Anal. Meth. in Geom., 2, 159-175, 1978.

125. NAGTEGALL, J.C., PARKS, D.M. and RICE, J.R., "On Numerically Accurate Finite Element Solutions in the Fully Plastic Range", Comp. Meth. Appl. Mech. Engng., 4, pp. 153-177, 1974.

APPENDIX A

STRESS DETERMINATION AT BOUNDARY NODES

Let us consider a boundary node "S_1" connected with two straight boundary elements, "j" and "j-1", for a two-dimensional body as shown in figure (A.1). In order to compute the stress tensor at node "S_1", two equations can be obtained using the traction-stress realtion (eq. 3.2.8) for plane strain or stress case as follows,

$$\bar{\sigma}_{12} = -\bar{p}_1$$

$$\bar{\sigma}_{22} = -\bar{p}_2 \qquad (A.1)$$

where the bar indicates the local system of coordinates.

For computing the third plane component of the stress tensor we have to use equation (3.2.17) in the following form,

$$\bar{\sigma}_{11} = \frac{2G}{1-\nu}\bar{\varepsilon}_{11} + \frac{\nu}{1-\nu}(\bar{\sigma}_{22}+\bar{\sigma}^o_{22}) - \bar{\sigma}^o_{11} \qquad (A.2)$$

in which $\bar{\varepsilon}_{11}$ is the strain component in the $\bar{X}_1$ direction.

Using linear interpolation functions, ϕ^1 and ϕ^2, the displacements along the element "j" are given by,

$$\bar{u}_i = \phi^1 \bar{u}_i^{(1)} + \phi^2 \bar{u}_i^{(2)} \qquad (A.3)$$

Expressing ϕ^1 and ϕ^2 in terms of the local coordinates, equation (A.3) becomes,

$$\bar{u}_1 = \frac{\bar{u}_1^{(1)} - \bar{u}_1^{(2)}}{2} + \frac{\bar{x}_1}{\ell}(\bar{u}_1^{(2)} - \bar{u}_1^{(1)}) \qquad (A.4)$$

Then, the strain component is obtained as follows,

$$\bar{\varepsilon}_{11} = \frac{\partial \bar{u}_1}{\partial \bar{x}_1} = (\bar{u}_1^{(2)} - \bar{u}_1^{(1)})/\ell \qquad (A.5)$$

Substituting (A.5) and (A.1) into (A.2) gives

$$\bar{\sigma}_{11} = \frac{2G}{1-\nu}(\bar{u}_1^{(2)} - \bar{u}_1^{(1)})/\ell - \frac{\nu}{1-\nu}\bar{p}_2 - \bar{\sigma}_{11}^o + \frac{\nu}{1-\nu}\bar{\sigma}_{22}^o \qquad (A.6)$$

As can be seen in equations (A.1) and (A.6), all stress components for a particular node are given as functions of both boundary values and initial stress components. Therefore, after expressing them in terms of the global system of coordinates, the same matrix form already employed for internal points can be adopted, i.e.,

$$\underline{\sigma} = - \underline{H}'\underline{U} + \underline{G}'\underline{P} + \underline{E}'\underline{\sigma}^o \qquad (A.7)$$

It is also important to notice that a convenient procedure must be chosen to take into account influences of both elements "j" and "j-1" in the stress values at the boundary node "S_1". For the applications presented in this work, the coefficients from adjacent elements have been averaged in order to assemble the matrices $\underline{H}'$, $\underline{S}'$ and $\underline{E}'$.

The shear stresses at a boundary node for the anti-plane case can also be obtained in a similar way. As only two shear stress values are defined for this case, the following relation between tractions, stresses and deformation at a boundary node can be defined,

$$\bar{\sigma}_{23} = - \bar{p}_3 \qquad (A.8)$$

$$\bar{\sigma}_{13} = G\bar{\varepsilon}_{13} - \bar{\sigma}_{13}^o \qquad (A.9)$$

As has been shown for the plane case, these expressions can also be written in a matrix form, as indicated in equation (A.7).

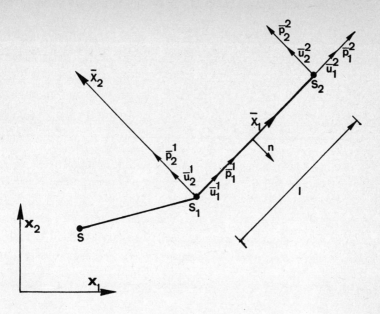

Figure A1. Local System of Coordinates.

APPENDIX B

SOME BASIC EXPRESSIONS FOR TWO-DIMENSIONAL
PLASTICITY AND VISCOPLASTICITY PROBLEMS

The formulation of the stress-strain relation for post-yield conditions presented in chapter 8 required the definition of tensors $a_{k\ell}$ and d_{ij}. The first was obtained from the derivatives of the yield function (F or G), with respect to the stress tensor components, i.e.,

$$a_{k\ell} = \frac{\partial}{\partial \sigma_{k\ell}} F \qquad (B.1)$$

The tensor d_{ij} was computed by multiplying $a_{k\ell}$ by the elastic compliances as follows,

$$d_{ij} = C_{ijk\ell} a_{k\ell} \qquad (B.2)$$

For both elastoplastic and elasto/viscoplastic techniques it is convenient to write the tensors $a_{k\ell}$ and d_{ij} in vectorial forms as follows,

$$\underset{\sim}{a} = \begin{bmatrix} a_{11} \\ a_{22} \\ 2a_{12} \\ a_{33} \\ 2a_{13} \\ 2a_{23} \end{bmatrix} \begin{matrix} \updownarrow \text{plane strain} \\ \text{or} \\ \text{plane stress} \end{matrix} \quad \begin{matrix} \updownarrow \text{complete} \\ \text{plain} \\ \text{strain} \end{matrix} \qquad (B.3)$$

in which coefficients related to shear stresses ($a_{k\ell}$, $k \neq \ell$) are multiplied by 2 in order to take into account the symmetric values, $a_{k\ell}$ and $a_{\ell k}$.

For the determination of the vector $\underline{d}$ it is convenient to express the elastic matrix in its explicit form, as follows,

$$\underline{C} = \frac{2G(1-\nu)}{(1-2\nu)} \begin{bmatrix} 1 & \frac{\nu}{1-\nu} & 0 & \frac{\nu}{1-\nu} & 0 & 0 \\ \frac{\nu}{1-\nu} & 1 & 0 & \frac{\nu}{1-\nu} & 0 & 0 \\ 0 & 0 & \frac{1-2\nu}{2(1-\nu)} & 0 & 0 & 0 \\ 0 & 0 & 0 & 1 & \frac{1-2\nu}{2(1-\nu)} & 0 \\ 0 & 0 & 0 & 0 & 0 & \frac{1-2\nu}{2(1-\nu)} \end{bmatrix} \quad (B.4)$$

$\longleftarrow$ for plane strain problems $\longrightarrow$

$\longleftarrow$ for complete plane strain problems $\longrightarrow$

and

$$\underline{C} = \frac{2G}{(1-\nu)} \begin{bmatrix} 1 & \nu & 0 & 0 \\ \nu & 1 & 0 & 0 \\ 0 & 0 & \frac{1-\nu}{2} & 0 \\ 0 & 0 & 0 & 0 \end{bmatrix} \quad (B.5)$$

for plane stress problems.

As indicated in equation (B.2), the vector form of d_{ij} can be obtained by multiplying the matrix $\underline{C}$ by the vector $\underline{a}$ i.e.,

$$\underset{\sim}{d} = \underset{\sim}{C}\underset{\sim}{a} \tag{B.6}$$

After performing the product one has,

$$\underset{\sim}{d} = \begin{bmatrix} a_{11} + M_1 \\ a_{22} + M_1 \\ a_{12} \\ a_{33} + M_1 \\ \hline a_{13} \\ a_{23} \end{bmatrix} \begin{array}{c} \uparrow \\ \text{for} \\ \text{plane} \\ \text{strain} \\ \downarrow \end{array} \begin{array}{c} \uparrow \\ \text{for} \\ \text{complete} \\ \text{plane} \\ \text{strain} \\ \downarrow \end{array} \tag{B.7}$$

in which

$$M_1 = \frac{\nu}{1-2\nu}(a_{11}+a_{22}+a_{33}) \tag{B.8}$$

and

$$\underset{\sim}{d} = \begin{bmatrix} a_{11} + M_2 \\ d_{22} + M_2 \\ a_{12} \\ 0 \end{bmatrix} \qquad M_2 = \frac{\bar{\nu}}{1-2\bar{\nu}}(a_{11}+a_{22}) \tag{B.9}$$

for plane stress problems.

The vectors $\underset{\sim}{a}$ and $\underset{\sim}{d}$ presented above can be introduced into the boundary formulation in order to improve the computing time spent by the calculation of both plastic stress increment (eq. 8.3.35) and viscoplastic stress rate (eq. 9.3.3).